NANOTECHNOLOGY SCIE

ION-SYNTHESIS OF SILVER NANOPARTICLES AND THEIR OPTICAL PROPERTIES

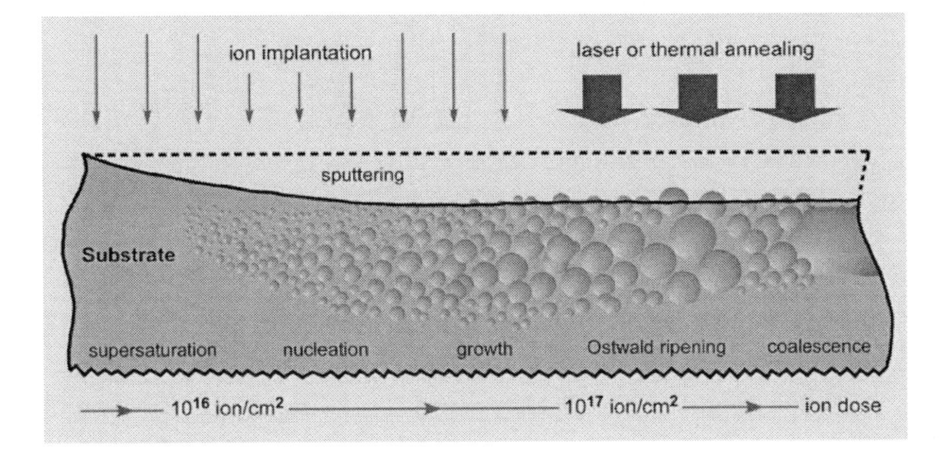

NANOTECHNOLOGY SCIENCE AND TECHNOLOGY

Additional books in this series can be found on Nova's website under the Series tab.

Additional E-books in this series can be found on Nova's website under the E-book tab.

NANOTECHNOLOGY SCIENCE AND TECHNOLOGY

ION-SYNTHESIS OF SILVER NANOPARTICLES AND THEIR OPTICAL PROPERTIES

ANDREY L. STEPANOV*

Kazan Physical-Technical Institute,
Russian Academy of Sciences,
Kazan, Russian Federation
Kazan State University, Kazan, Russian Federation

Nova Science Publishers, Inc.

New York

* e-mail: aanstep@gmail.com anstep@kfti.knc.ru

For permission to use material from this book please contact us:
Telephone 631-231-7269; Fax 631-231-8175
Web Site: http://www.novapublishers.com

Additional color graphics may be available in the e-book version of this book.

LIBRARY OF CONGRESS CATALOGING-IN-PUBLICATION DATA

Stepanov, Andrey L.
Ion-synthesis of silver nanoparticles and their optical properties /
Andrey L. Stepanov.
p. cm.
Includes index.
ISBN 978-1-61668-862-2 (softcover)
1. Silver--Optical properties. 2. Nanoparticles--Optical properties. 3. Dielectrics--Optical properties. 4. Ion implantation. 5. Ionic structure.
I. Title.
TA480.S5S74 2009
620.1'8923--dc22

2010016672

Published by Nova Science Publishers, Inc. ✝ New York

CONTENTS

ABSTRACT

Recent results on ion-synthesis by low-energy implantation and optical properties of silver nanoparticles in various dielectrics (glasses and polymers) and on the interaction of high-power laser pulses with such composite materials are reviewed. One of the features of composites prepared by the low-energy ion implantation is the growth of metal particles with a wide-size distribution in the thin depth from the irradiated substrate surface. This leads to specific optical properties of implanted materials, partially to difference in reflection measured form implanted and rear face of samples. The excimer laser pulse modification of silver nanoparticles fabricated in silicate glasses is considered. Pulsed laser irradiation makes it possible to modify such composite layer, improving the uniformity in the size distribution of the nanoparticles.The optical absorption of silver nanoparticles fabricated in polymer is also analysed. Unusual weak and broad plasmon resonance spectra of the nanoparticles are studied in the frame of the carbonisation of ion-irradiated polymer. Based on the Mie theory, optical extinction spectra for metal particles in the polymer and carbon matrices are simulated and compared with particle spectra for complex silver core–carbon shell nanoparticles. A new experimental data on nonlinear optical properties of synthesised silver nanoparticles are also presented.

INTRODUCTION

Nanomaterials are cornerstones of nanoscience and nanotechnology. The relevant feature size of nanomaterial components is on the order of a few to a few hundreds of nm. At the fundamental level, there is a real need to better understand the properties of materials on the nanoscale level. At the technological front, there is a strong demand to develop new techniques to fabricate and measure the properties of nanomaterials and relevant devices. Significant advancement was made over the last decades in both fronts. It was demonstrated that materials at the nanoscale have unique physical and chemical properties compared to their bulk counterparts and these properties are highly promising for a variety of technological applications. One of the most fascinating and useful aspects of nanomaterials is their optical properties. Applications based on optical properties of nanomaterials include optical detectors, laser, sensor, imaging, display, solar cell, photocatalysis, photoelectrochemistry, and biomedicine [1]. Among variety of nanomaterial a most fascinating ones are composite materials containing metallic nanoparticles (MNPs) which now considered as a basis for designing new photonic media for optoelectronics and nonlinear optics [2]. Simultaneously, with the search for and development of modern technologies intended for nanoparticle synthesis, substantial practical attention was devoted to designing techniques for controlling the MNP size. This is caused by the fact that the properties of MNPs, such as the quantum size effect, single-electron conduction, etc., which are required for various applications, take place up to a certain MNP size. An example of their application in optoelectronics is a prototype of integrated electronic circuit - chip that combines metallic wires as conductors of electric signals with fibers as guides of optical signals. In practice, light guides are frequently made of synthetic sapphire or siliconoxide, which are deposited on or buried in semiconductor substrates. In this case, electrooptic emitters and that accomplish electric-to-optic signal conversion are fabricated inside the dielectric layer. This light signal from a microlaser is focused in a light guide and then transmitted through the optoelectronic chip to a high-speed photodetector, which converts the photon flux to the flux of electrons. It is expected that light guides used instead of metallic conductors will improve the data rate by at least two orders of magnitude. Moreover, there is good reason to believe that optical guide elements will reduce the energy consumption and heat dissipation, since metallic or semiconductor components of the circuits may be replaced by dielectric ones in this case. Prototype optoelectronic chips currently available are capable of handling data streams with a rate of 1 Gbit/s, with improvement until 10 Gbit/s in future. Key elements of dielectric waveguides used for light propagation are nonlinear optical switches, which must provide conversion of laser signal for pulse duration as short as pico- or femtoseconds. The nonlinear optical properties of MNP-containing dielectrics stem from the dependence of their refractive index and nonlinear absorption on incident light intensity [2, 3]. This effect is associated with MNPs, which exhibit an enhancement of local electromagnetic field in a composite and, as consequence, a high value of the third order nonlinear susceptibility when exposed to ultra-short laser pulses. Therefore, such MNP-containing dielectric materials may be used to advantage in integrated optoelectronic devices [4]. A local field enhancement in MNPs stimulates a strong linear optical absorption called as surface plasmon resonance (SPR). The electron transitions responsible for plasmon absorption in MNPs also cause a generation of an optical nonlinearity of a composite in the same spectral range. As a result, the manifestation of nonlinear optical properties is most efficient for wavelengths near the position of a SPR maximum. In practice, to reach the strong linear absorption of a composite in the SPR spectral region, attempts are made to increase the concentration (filling factor) of MNPs. Systems with a higher filling factor offer a higher nonlinear susceptibility, when all other parameters of composites being the same. Usually noble metals and copper are used to fabricate nonlinear optical materials with high values of third order susceptibility.

Small size alters the electronic structure of MNPs. This provides greater pumping efficiency and lower overall threshold for applications in optical switching. The potential advantages of MNP composites as photonic materials are substantial improvement in the signal switching speed up to 100 GHz repetition frequencies are expected in communication and computing systems of the 21st century[3]. Fig. 1 compares in graphical formthe switching speeds and switching energies of a number of electronic and optical materials and devices (adapted from [3]). Within the broad range on parameters covered by "conventional semiconductor microelectronics", current metal-oxide-semiconductor field-effect transistor devices made in Si have low switching energies, but switching time in the nanosecond range. Photonic devices based on multiple quantum well (MQW) structures – SEED and GaAs MQW devices and Fabry-Perot (FP) cavities based on ferroelectric such as lithium niobate – have extremely low switching energies in comparison to MNPs, but relatively slow switching speed. As seen, MNPs fit into the area of current semiconductor electronics: they have very rapid switching times, as low as picoseconds and femtoseconds. Unfortunately, so far, still relatively little attention has been paid to the practical problems associated with the realization of electo-optical device structures on silicon platform— the analog of building up transistor structures (sources, gates, electrodes) for microelectronic applications.

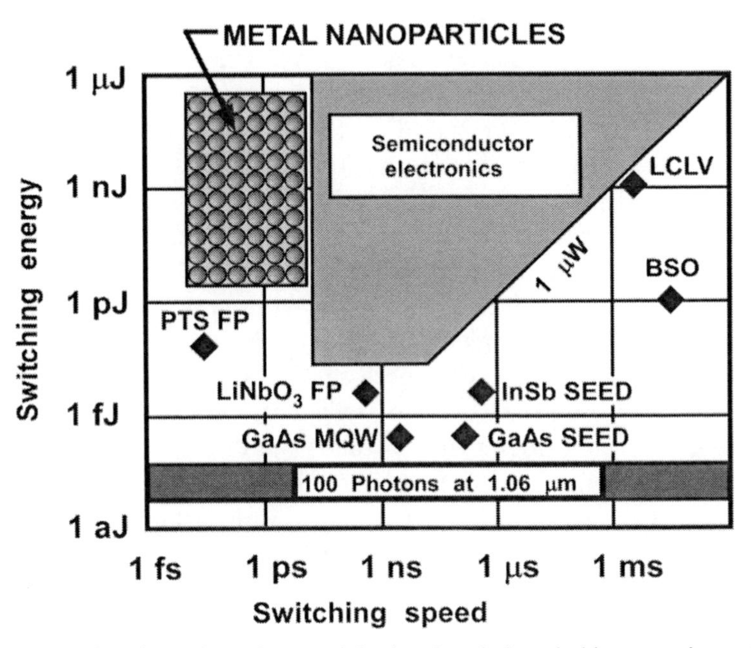

Figure 1. Plot of various photonic materials showing their switching energies and switching speeds.

There are variety ways to synthesis MNPs in dielectrics, such as magnetron sputtering, the convective method, ion exchange, sol–gel deposition, etc. One of the most promising enhanced fabrication methods is ion implantation [5-9] because it allows reaching a high metal filling factor in an irradiated matrix beyond the equilibrium limit of metal solubility and provides controllable synthesis of MNPs at various depths under the substrate surface. Nearly any metal–dielectric composition may be produced using ion implantation. This method allows for strict control of the doping ion beam position on the sample surface with implant dose as, for example, in the case of electron- and ion-beam lithography. Today, ion implantation is widely used in industrial semiconductor chip fabrication. Therefore, the combination of MNP-containing dielectrics with semiconductor substrates by same technological approach as ion implantation could be reached quite effective. Moreover, ion implantation can be applied for different steps in optoelectronic material fabrication such as creation of optical waveguides by implantation with rear gas ions (H^+, He^+ etc.) [9], a designing of electric-to-optic signal convectors and microlaser by irradiation of dialectics waveguides with rear metal ions (Er^+, Eu^+ etc.) [9, 10] and a synthesis of MNPs (Fig. 2).

The history of MNP synthesis in dielectrics by ion implantation dates back to 1973, when a team of researchers at the Lyons University in France [11, 12] pioneered this method to create particles of various metals (sodium, calcium, etc.) in LiF and MgO ionic crystals. Later, ion-synthesis of noble nanoparticles was firstly done in study of Au- and Ag-irradiated lithia-alumina-silica glasses [13, 14]. Development was expanded from the metal implants to the use of many ions and the active formation of compounds, including metal alloys and totally different composition precipitate inclusions. In ion implantation practice, MNPs were fabricated in various materials, such as polymers, glass, artificial crystals, and minerals [15, 16]. By implantation, one can produce almost any metal–dielectric composite materials, as follows from Table 1, which gives a comprehensive list of references of various dielectrics with implanted silver nanoparticles with conditions for their fabrications.

The book focuses on recent advantages in fabrication of silver nanoparticles by low-energy in implantation in various inorganic matrixes [17-161]. Also, some examples of nonlinear optical repose in such composites are presented and discussed.

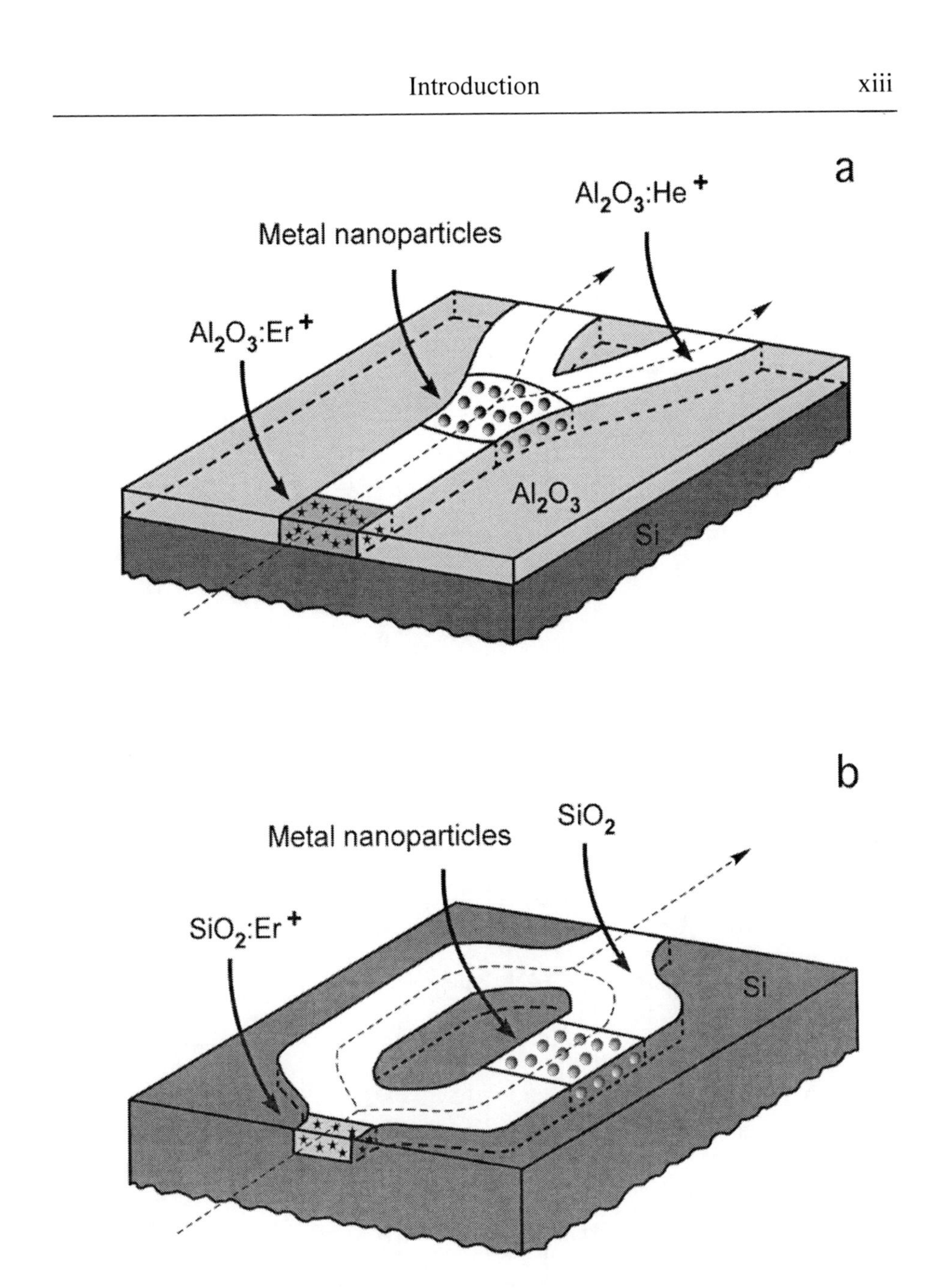

Figure 2.A prototype of optoelectronic chip with a dielectric waveguide combined with silicone substrate. Ion implantation can be applied to fabricate selective area doped by rear metal ions (marked by stars) to work as microlaser and to illuminate in waveguide, created by rear-gas ion radiation with MNPs to form an optical switcher.

Table 1. Types of dielectric inorganic matrix with silver nanoparticles synthesized by ion implantation combined in some cases with post-implantation heat treatment. Abbreviations – $2Ag_2O{\cdot}3Na_2O{\cdot}25ZnO{\cdot}70TeO_2$(ANZT glass), alkali-borosilicate glass (ABSG), borosilicate Pyrex glass (BPYR), soda-lime silicate glass (SLSG),yttria stabilized cubic zirconia (YSZ);optical reflection (OR), optical absorption (OA), transmission electron microscopy (TEM), TEM cross-section (TEM-CS), high resolution TEM (HRTEM), scanning transmission electron microscopy (STEM), conductivity measurements (CM), atom force microscopy (AFM), optical microscopy (OM), selected area electron diffraction (SAED), energy dispersiveX-ray spectrometry(EDS), high-resolution X-ray diffraction (XRD), Z-scan, RZ-scan by reflection, degenerate four wave mixing (DFWM); room temperature (RT).

Matrix type	Ion energy, keV	Ion dose,ion/cm^2	Current density,μA/cм^2	Matrix temperature,°C	Post-implantation heat treatment	Methodsof particle detection	Authors
Al_2O_3 crystal <1010>	50 360	$4.0{\cdot}10^{16}$ $5.0{\cdot}10^{16}$ $8.0{\cdot}10^{16}$	1-5	77 K 300	Annealing in air at 650°C, 30 min.	OA	Rahmani *et al.* 1988 [17] 1989 [18]
Al_2O_3 crystal <0001>	$1.8{\cdot}10^3$	$1.2{\cdot}10^{17}$		77 K	Annealing in airat 1100°C, 2 h	OA, DFWM	White *et al.* 1993 [19]
Al_2O_3 crystal	$1.5{\cdot}10^3$	$8.0{\cdot}10^{16}$	2<		Annealing in airat 500°C, 1 h	OA	Ila *et al.* 1998 [20]
Al_2O_3 crystal	25 30	$(0.2\text{-}2.0){\cdot}10^{17}$				OA, OR, TEM	Steiner *et al.* 1998 [21]
Al_2O_3 crystal	30	$3.8{\cdot}10^{17}$	3, 6, 10	RT		OA RZ-scan	Ganeev *et al.* 2005 [22] 2006 [23]
Al_2O_3 crystal	160	$(0.1\text{-}1.0){\cdot}10^{17}$				OA	Marques *et al.* 2006 [24]
ABSG, BPYR glass	270	$1.5{\cdot}10^{16}$				OA	Mazzoldi *et al.* 1993 [25]
MgO crystal (100)	180	(0.5-1.0)·1017	1	RT	Some samples annealed in vacuum at 25-1500°C	OA TEM	Abouchacra and Serughetti 1986 [26] Fuchs et al. 1988 [27]

Table 1. (Continued)

Matrix type	Ion energy, keV	Ion dose,ion/cm2	Current density,μA/cм2	Matrix temperature,°C	Post-implantation heat treatment	Methodsof particle detection	Authors
MgO crystal (100)	$1.5{\cdot}10^3$	$1.2{\cdot}10^{17}$	2-3	27 C	Annealing in air at 550 and 1100°C	OA TEM	Qian *et al.* 1997 [28] Zimmerman *et al.* 1997 [29]
MgO crystal (100)	600	$1.0{\cdot}10^{16}$		RT	Annealing in air at 1200°C, 22 h	OA, XRD TEM-CS	van Huis *et al.* 2002 [30]
MgO crystal (100)	200	$2.0{\cdot}10^{17}$	2	RT	Some samples annealed in air, Ar, O_2 or 70%N_2 + 30%H_2, at 300-900°C, 1 h	OA, SAED TEM-CS	Xiao *et al.* 2008 [31]
$MgOP_2O_5$ glass	150	$(0.1\text{-}1.0){\cdot}10^{17}$	0.5-3	RT		OA	Matsunami and Hosono 1993 [32]
Lithia-alumina-silica glass	275-285	$1.0{\cdot}10^{16}$	1-2	300		OA	Arnold and Borders 1976 [14]
$LiNbO_3$ crystal	50 360	$(4.0\text{-}0.8){\cdot}10^{16}$	1-4	77 K 300	Annealing in air at 250-650°C, 30 min	OA	Rahmani *et al.* 1988 [17] 1989 [18]
$LiNbO_3$ crystal	20, 25 $3{\cdot}10^3$ $4.2{\cdot}10^3$	$(0.5\text{-}8.0){\cdot}10^{16}$		RT	Some samples annealed in air at 200-600°C, 1-3 h	OA OM X-ray	Deying *et al.* 1994 [33] Shang *et al.* 1996 [34] Saito and Kitahara 2000 [35] Fujita *et al.* 1994 [36]
$LiNbO_3$ crystal	160 $1.5{\cdot}10^3$	$2.0{\cdot}10^{16}$ $4.0{\cdot}10^{16}$ $1.7{\cdot}10^{17}$		RT 500	Some samples annealed in air at 500-800°C, 1 h	OA Z-scan TEM TEM-CS	Sarkisov *et al.* 1998 [27-40] 1999 [41] 2000 [42] Williams *et al.* 1998 [43, 44] 1999 [45]
$LiNbO_3$ crystal	$1.5{\cdot}10^3$	$2.0{\cdot}10^{16}$		300	Annealing in Ar gas at 100-1100°C, 30 min	OA TEM TEM-CS	Amolo *et al.* 2006 [46]
SiO2 crystal	200	(2.3-9.0)·1016	1-5	77 300	Annealing in air at 300-500°C, 30 min	OA	Rahmani and Townsend1989 [18]

Table 1. (Continued)

Matrix type	Ion energy, keV	Ion dose,ion/cm2	Current density,μA/cм2	Matrix temperature,°C	Post-implantation heat treatment	Methodsof particle detection	Authors
SiO_2	65 130 270	$(1.5\text{-}5.0)\cdot 10^{16}$	1 1.5	300	Some samples annealed in air or 4 % H_2 gas	OA TEM EXAFS	Mazzoldi *et al.* 1993 [47] Mazzoldi *et al.* 2005 [48] Mazzoldi *et al.* 2007 [49] Antonello *et al.* 1998 [50] Battaglin *et al.* 1998 [51] 2001 [52] Bertoncello *et al.* 1998 [53] Caccavale 1998 [54] Cattaruzza *et al.* 1999 [55] Gonella *et al.* 1999 [56]
SiO_2	150	$(0.1\text{-}6.0)\cdot 10^{17}$	1.5-14	300		TEM-CS, OA	Matsunami and Hosono 1993 [57]
SiO_2	305	$(3.0\text{-}9.0)\cdot\cdot 10^{16}$	2	0		OA TEM X-ray	Magruder III *et al.* 1995 [58] 1996 [59] 2009 [60] Anderson *et al.* 1996 [61] 1997 [62] 1998 [63] 2000 [64] Zuhr *et al.* 1998 [65]
SiO_2	20-58 130	$(0.4\text{-}2.0)\cdot 10^{17}$	0.6	300		AFM	Pham *et al.* 1997 [66]
SiO_2 crystal	200	$(2.3\text{-}9.0)\cdot 10^{16}$		300		OA TEM	Liu *et al.* 1998 [67-69]
SiO_2	$1.5\cdot 10^3$	$2.0\cdot 10^{16}$ $4.0\cdot 10^{16}$ $1.4\cdot 10^{17}$	2		Annealing in Ar gas at 500-1000°C, 1 h	OA Z-scan	Ila *et al.* 1998 [70]
SiO_2	60	$4.0\cdot 10^{16}$	10	RT		OA, AFM	Stepanov *et al.* 2000 [72] 2003 [73]

Table 1. (Continued)

Matrix type	Ion energy, keV	Ion dose,ion/cm2	Current density,μA/cм2	Matrix temperature,°C	Post-implantation heat treatment	Methodsof particle detection	Authors
SiO2	65	5.0·1016				X-ray EXAFS	D'Acapito and Zontone 1999 [71]
SiO_2	43 90 150 200 300	$(0.06\text{-}2.0)\cdot10^{17}$	0.8-2.5	300	Some samples annealed in air, Ar, O_2 or 70%N_2+ 30%H_2 at 300-800°C, 1 h	SAED OA, STEM TEM, EDS HRTEM TEM-CS Z-scan	Jiang *et al.* 2000 [74] Ren *et al.* 2004 [75, 76] 2005 [77-79] 2006 [80] 2007 [81] 2008 [82] 2009 [83] Liu *et al.* 2005 [84] Xiao *et al.* 2006 [85] 2007 [86, 87] Wang *et al.* 2007 [88] Wang *et al.* 2008 [89] Zhang *et al.* 2004 [90] Cai *et al.* 2008 [91] 2009 [92]
SiO_2 sol-gel film	5-100	$(5.0\text{-}6.0)\cdot10^{16}$	1.5-2.5	330 K		TEM HR-TEM XRD SAED	Armelao*et al.* 2002 [93]
SiO_2 on Si	10 30 40	$(1.0\text{-}5.0)\cdot10^{15}$ $(1.0\text{-}5.0)\cdot10^{16}$ $1.0\cdot10^{17}$	2	-	Annealing in Ar gas at 500-900°C, 1 h	OR, OR TEM-CS R TEM HR-TEM	Ishikawa *et al.* 2002 [94] 2009 [95] Tsuji *et al.* 2002 [96, 97] 2003 [98] 2004 [99] 2005 [100] Arai *et al.* 2003 [101] 2005 [102] 2006 [103] 2007 [104, 105]

Table 1. (Continued)

Matrix type	Ion energy, keV	Ion dose,ion/cm2	Current density,μA/см2	Matrix temperature,°C	Post-implantation heat treatment	Methodsof particle detection	Authors
SiO_2	$2.0 \cdot 10^3$	$(0.4\text{-}1.0) \cdot 10^{17}$	2	RT	Some samples annealed in 50%N_2 + 50%H_2 gas or in air at 230-800°C, 1 h	OA TEM HRTEM	Roiz *et al.* 2004 [106] Oliver *et al.* 2006 [107] Cheang-Wong *et al.* 2007 [108] Peña *et al.* 2007 [109] 2009 [110] Reyes-Esqueda *et al.* 2008 [111] 2009 [112] Rodrigues-Iglesias *et al.* 2008 [113] 2009 [114] Rangel-Rojo *et al.* 2009 [115]
SiO_2 on Si	40	$0.3 \cdot 10^{15}$		RT	Annealing in vacuumat 550°C, 20 min	TEM	Romanyuk *et al.* 2006 [116]
SiO_2	60	$(0.3\text{-}1.0) \cdot 10^{17}$	3	RT		OA Z-scan	Takeda *et al.* 2006 [117]
SiO_2	32-40 $1.7 \cdot 10^3$ $2.4 \cdot 10^3$	$(0.1\text{-}1.0) \cdot 10^{17}$	3-5	RT	Some samples annealed in air at 500°C, 1 h	OA TEM Z-scan	Joseph *et al.* 2007 [118, 119] Sahu *et al.* 2009 [120]
SiO_2	0.65 1.5 3 keV	$(1.2\text{-}4.7) \cdot 10^{15}$	3-5	RT		OA, OR TEM HRTEM TEM-CS	Carles *et al.* 2009 [121]
SiO_2	200	$(0.1\text{-}2.0) \cdot 10^{17}$	< 2.5	-		OA TEM TEM-CS Z-scan	Wang *et al.* 2009 [122, 123]
SiO_2+TiO	305	$6.0 \cdot 10^{16}$	7	-		OA TEM-CS	Magruder III *et al.* 2007 [124]

Table 1. (Continued)

Matrix type	Ion energy, keV	Ion dose,ion/cm2	Current density,μA/cм2	Matrix temperature,°C	Post-implantation heat treatment	Methodsof particle detection	Authors
Si3N4	20 130	4.0·1016	0.6	300		AFM	Pham et al. 1997 [66]
BPYR glass	270	1.5·1016				OA	Mazzoldi et al. 1993 [25]
Soda-lime glass	60	$2.0 \cdot 10^{16}$ $4.0 \cdot 10^{16}$	-	300	-	OR TEM-CS	Nistor *et al.* 1993 [125] Wood *et al.* 1993 [1126]
Soda-lime glass	200	$(0.5\text{-}4.0) \cdot 10^{16}$	0.5-2	RT 77 K	-	TEM TEM-CS OA	Dubiel *et al.* 1997 [127] 2000 [128] 2003 [129] 2008 [130] Seifert *et al.* 2009 [131]
Soda-lime glass	200	$(0.5\text{-}4.0) \cdot 10^{16}$	0.5-2	RT 77 K	-	TEM TEM-CS OA	Stepanov *et al.* 1998 [132] 1999 [133-135] 2000 [136-139] 2001 [140-142] 2002 [143-146] 2003 [147-149] 2004 [150, 151] 2005 [152] 2008 [153, 154] 2009 [155]
Ta_2O_5	80-130	$6.0 \cdot 10^{16}$	0.6 – 6.4	300		AFM	Pham *et al.* 1997 [66]
TiO_2 crystal	50 65	$(0.3\text{-}1.0) \cdot 10^{17}$	2	-	Annealing in Ar gas at >400°C, 1 h	OA TEM-CS	Tsuji *et al.* 2002 [156] 2003 [157]
TiO_2 Sol-gel films	30 65	$(0.1\text{-}0.5) \cdot 10^{17}$	2	RT	Annealing in Ar gas at 300-600°C, 1 h	OA	Tsuji *et al.* 2005 [158] 2006 [159]
YSZ	20 $1.5 \cdot 10^3$ $3.0 \cdot 10^3$	$(0.7\text{-}6.0) \cdot 10^{16}$	2	RT	Some samples annealed in air at 500-1000°C	OA	Saito *et al.* 2003 [160] Fujita *et al.* 2007 [161]

Chapter 1

DEPTH DISTRIBUTION AND DIFFUSION OF SILVER IONS DURING IMPLANTATION

The formation of MNPs resulting from ion implantation into dielectric substrates is complex, since there are a large number of influencing factors. A simple ion range estimate of the silver concentration can be computed, but this is only a precursor of processes involving diffusion and clustering and so simple simulations of the entire process are extremely challenging. Thus, the process should be divided into sub-processes with a time scale that resolves implantation, diffusion, nucleation and particle growth. The first step for consideration is the dependence the ion depth distribution caused by silver diffusion at different substrate temperatures. At simple consideration, during the irradiation, implanted ions leads to a depth distribution in the substrate, which has approximately a Gaussian shape, as described by range algorithms such as TRIM (SRIM) [162]. The diffusion equation of ion-implanted impurities is assumed to be expressed as [163]:

$$\frac{\partial N(x,t)}{\partial t} = D \cdot \frac{\partial^2 N(x,t)}{\partial x^2} + n(x,t), \qquad (1)$$

where $N(x,t)$ is the concentration of impurities, D is their diffusion coefficient, $n(x,t)$ is the generation rate of the impurities due to ion implantation, x is distance (depth) and t is a duration of implantation. The diffusion coefficient in Eq. 1 is assumed to be independent of the distance x in the following calculation. D will depend on the rate of vacancy formation and the pre-existing concentration of silver particles, which act as trapping sites.

Initially,the generation rate $n(x,t)$ is believed to be of a Gaussian form [9, 163] and is given by

$$n(x,t) = \frac{\Phi}{\Delta R_p (2\pi)^{1/2}} \exp\left[-\frac{1}{2}\left(\frac{x - R_p}{\Delta R_p} \right)^2 \right], \tag{2}$$

where Φ is the dose rate per unit area of impurity ions, R_p is the projected range of an implanted ion, ΔR_p is the projected range straggling.

Let us consider, as example, the $R_p$and ΔR_p corresponding to Ag-implantation into SLSG for different energies calculated by TRIM (SRIM) programme [162]. The concentration profiles for different implant temperatures of Ag ions in SLSG are given by solution of Eq.1 and 2 [164] as

$$N(x,t) = \Phi\sqrt{\frac{2Dt + \Delta R_p^2}{2\pi D^2}}\left(-\frac{\alpha^2}{4Dt + 2\Delta R_p^2} \right)$$

$$- \Phi\sqrt{\frac{\Delta R_p^2}{2\pi D^2}} \, exp\left(\frac{\alpha^2}{2\Delta R_p^2} \right) +$$

$$+ \frac{\alpha\Phi}{2D}\left(erfc \frac{\alpha}{\sqrt{4Dt + 2\Delta R_p^2}} - erfc \frac{\alpha}{\Delta R_p \sqrt{2}} \right) \tag{3}$$

where $\alpha = x - R_p$ and t is a duration of ion implantation. As seen from Eq.3, the diffusion coefficient, which is dependent on temperature, determines the shape of the concentration profile. For an estimate of the silver diffusion coefficient in SLSG the Arrhenius equation may be applied with values of 0.69 eV for the activation energy and $5.6 \cdot 10^{-5}$ cm^2/sec for frequency factor [165]. If these values are suggested to be time independent for a fixed temperature, then the results of concentration profile calculations for an applied dose rate of $5.58 \cdot 10^{13}$ ion/cm^2 and a 360 second duration implantation, which correspond to a total dose of $2 \cdot 10^{16}$ ion/cm^2, are presented in Fig. 3. As seen in the figure, increasing the temperature from 20 to 100°C and, consequently, increasing the Ag diffusion coefficient in the glass from $2.88 \cdot 10^{-17}$ to $2.66 \cdot 10^{-14}$ cm^2/s, leads to a broadening of the initial Gaussian concentration profile and a reduction of the concentration at the peak of the profile. This decreasing in concentration is most critical for samples implanted at low energy especially. Thus, the accumulation of implanted ions in the SLSG layer is strongly affected by the substrate temperature, and hence this in turn influences the rate and depth of the development of the conditions for reaching a sufficient impurity concentration for MNP nucleation and growth. Obviously, if the Ag mobility is rather high, there is no possibility for nanoparticle nucleation during a reasonable implant time. Such an inhibiting effect had been clearly seen in earlier experiments which recorded depth profiles by RBS measurements of the similar type of float glass implanted with Ag ions at substrate temperatures higher than ~180°C [166]. Note also that for this calculation it is assumed that the bulk glass temperature, and the local temperature within the implanted layer, are the same. In practice, the surface will be heated to a higher temperature than the bulk of the glass.

Figure 3 does not include the influence of diffusion limited by the appearance of metal particles in the implanted material. However, it was shown, both in an example of implantation of Ag ions into SiO_2 glass [57] that the impurity diffusion coefficient drops dramatically after MNP formation which act as traps for mobile ions. This suggests that the critical time for control of the nanoparticle spacing and nucleation is at the beginning of the implant, and therefore both the substrate temperature and ion beam currents during this initial phase are crucial. After particle formation has commenced, any changes, such as increasing temperature from beam heating or increases in ion current, will presumably have effects on the particle sizes, but less influence on the depth profile of the distribution. High temperature conditions in the initial stages will increase the impurity diffusion and so reduce the supersaturation, which is required for particle nucleation, hence nanoparticles may not form. These conclusions are important as they emphasize that there is a need to control the temperature and ion beam current density throughout the implant. Many experimentalists fail to do this, but instead allow the temperature to rise from the beam heating. In some cases the initial dose is provided at a low current density in order to avoid surface charging, and hence changes in the ion beam energy. Once some implantation has occurred, the surface conductivity is increased and, hence, the beam currents can be raised. The foregoing conclusions suggest both situations influence the nanoparticle sizes and their depth distributions.

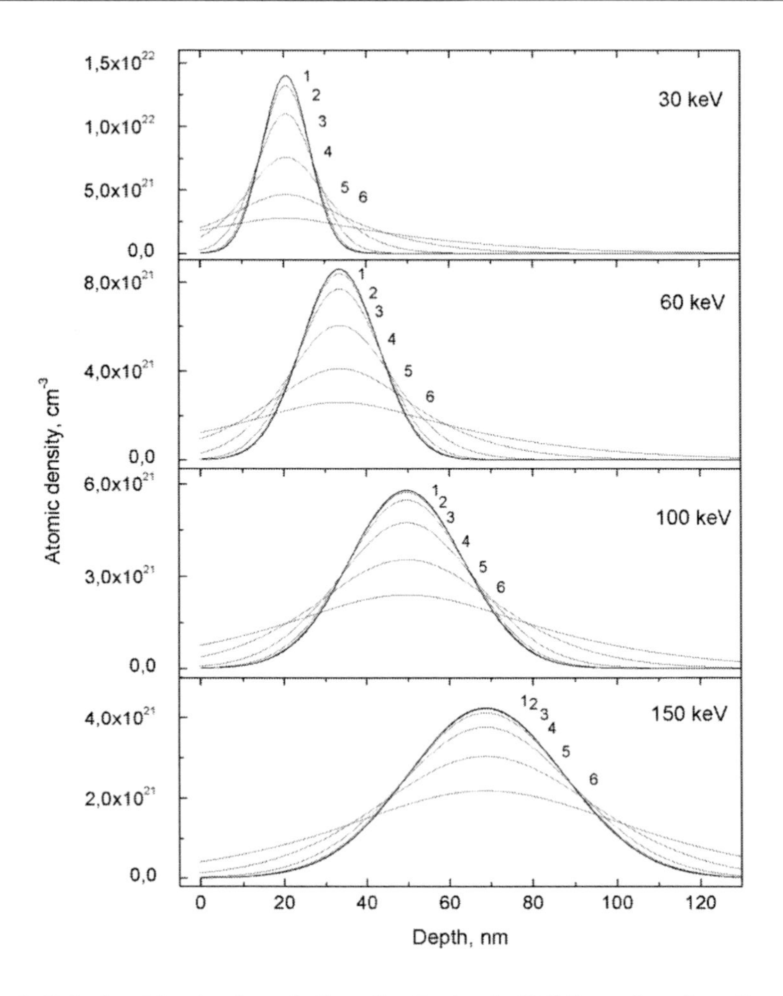

Figure 3. Calculated ion implanted silver distribution in SLSG as a function of energy, after taking into account the impurity diffusion in dependence on substrate temperature: 1.- TRIM distribution; 2.- 20; 3.-40; 4.- 60; 5.- 80; 6.-100°C. The concentration profiles correspond to the $R_p$and ΔR_p of 20.6 and 5.7 nm (30 keV), 33.6 and 9.3 nm (60 keV), 49.6 and 13.8 nm (100 keV), 68.6 and 18.8 nm (150 keV), respectively [134].

Chapter 2

Depth Distribution of Silver Ions in Dependence on Irradiation Energy and Surface Sputtering Effects

As was noticed, in first approximation, implantation of ions leads to a depth distribution in the substrate, which is approximately Gaussian as described by range algorithms such as TRIM (SRIM) [162]. However, the interaction of implanted ions with the substrate leads to ejection of ions and neutrals from the surface [9, 163]. This sputtering yield is a function of the incoming ion energy, dose and the masses of the ion and target atoms. Figs. 4 and 5 show the calculated thickness of the sputtered layers for float glass, and the corresponding TRIM concentration profiles of the Ag-ion implantation [167]. Here, secondary features such as alterations in range with time dependent compositions after sputtering (and diffusion) was ignored. Nerveless, these figures demonstrate that for 60 keV Ag-implantation the depth concentration in the float glass differs from a Gaussian profile, and have a maximum concentration near the surface.

To take into account the alterations in range by dose effect changes in composition, new simulations, using a dynamic computer code DYNA [168, 169], based on binary collision approximations in intermixed layer formations and sputtering processes, were applied for Ag ion implantation intoamorphous insulators: SiO_2, Al_2O_3 and SLSG [137]. To include a change of the near-surface layer composition due to cascade atom mixing into a concentration profile calculation, the volume of atoms has to be initially estimated, and was determined here, from the element densities or interatomic separations in the substrates. The sputtering yields at normal ion incidence are dependent on the energy of the metal-ion implantation and were separately calculated using the SRIM-2000 (TRIM) programme [162] with the corresponding binding, surface and lattice energies for amorphous SiO_2, Al_2O_3 and SLSG. The elemental concentrations for ion energies of 30, 60 and 100 keV have been obtained at doses of 0.1, 0.3, 0.6 and $1 \cdot 10^{16}$ ion/cm^2. The dose step in the calculations was $5 \cdot 10^{14}$ ion/cm^2.

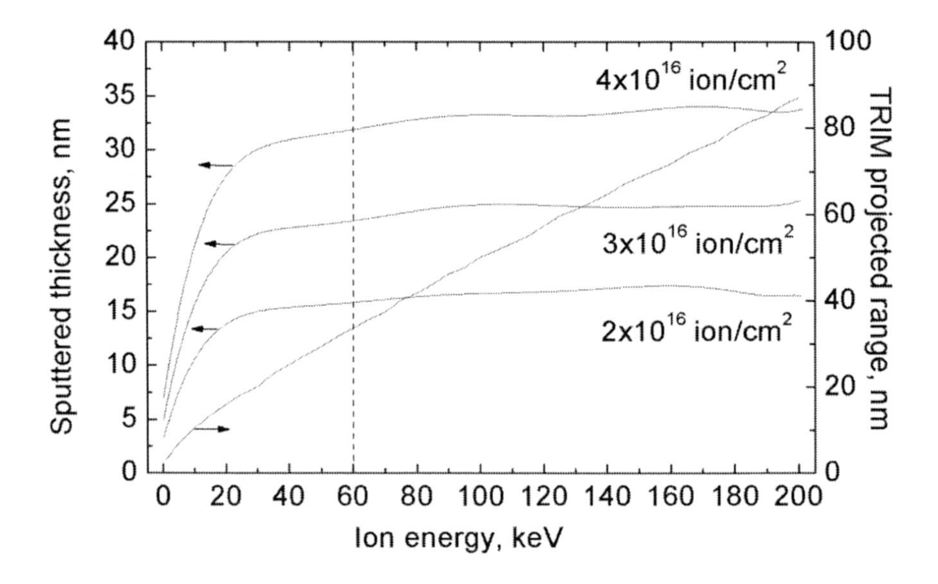

Figure 4. The calculated dependence of the Gaussian maximum in the depth, excluding sputtering (right hand scale), and thickness of the surface sputtered float glass layer (left hand scale) for Ag implanted into SLSG [167].

The results of DYNA calculations for Ag ion implantation into different dielectrics are presented in Fig. 6 [137]. Curves marked “TRIM” in these figures correspond to typical statistical TRIM calculations, which produce the Gaussian impurity distributions. Other curves 1-4 show the DYNA concentration profiles simulated for doses of 0.1, 0.3, 0.6 and $1 \cdot 10^{16}$ ion/cm^2. As shown here the peak position of the DYNA profiles appear closer to the implanted surface than the symmetrical TRIM curve. Also, the shapes of DYNA curves become asymmetrical when the dose exceeds a critical value.

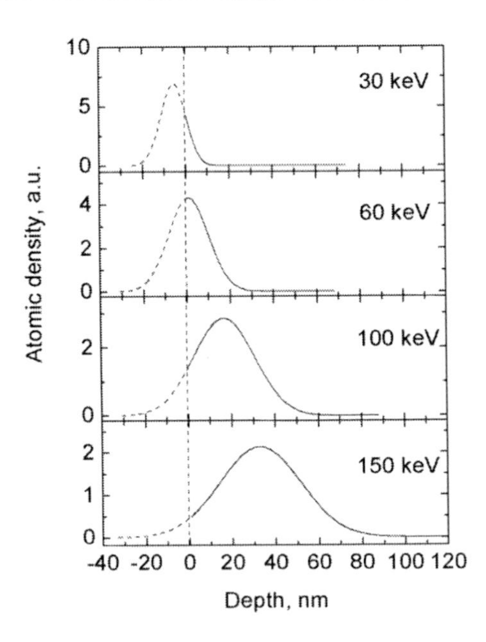

Figure 5. Calculated ion implanted silver distribution as a function of energy after taking into account the sputtering yield. The vertical dashed line indicates the final surface position, and the left part of the ion distribution from this point shows the sputtered ion portion [167].

For the cases of higher energy (60 – 100 keV) implantation in it is possible to see a dynamic development of the concentration profile during the time of accumulation of implanted ions in the substrates. At the start of the implantation the impurity distribution matches the TRIM curve. As is known, high dose irradiation can, in principle, alter or limit the ultimate concentrations attainable, because of some competition between the sputtering process, and change of both the composition and density of the surface substrate layer by introduction of ions and intermixing with volume atoms. During ion implantation, the sputtering process removes both target and implanted ions. Eventually, an equilibrium condition (steady state) may be reached, where as many implanted atoms are removed by sputtering as are replenished by implantation. The depth distribution of implanted atoms under this condition typically has a maximum at the surface and falls off over a distance comparable to the initial ion range. As seen in Fig. 6 this competition for the case of Ag ion implantation into dielectrics, leads to a shift of the concentration profile to the surface with increasing dose. Thus the profiles become very asymmetrical.

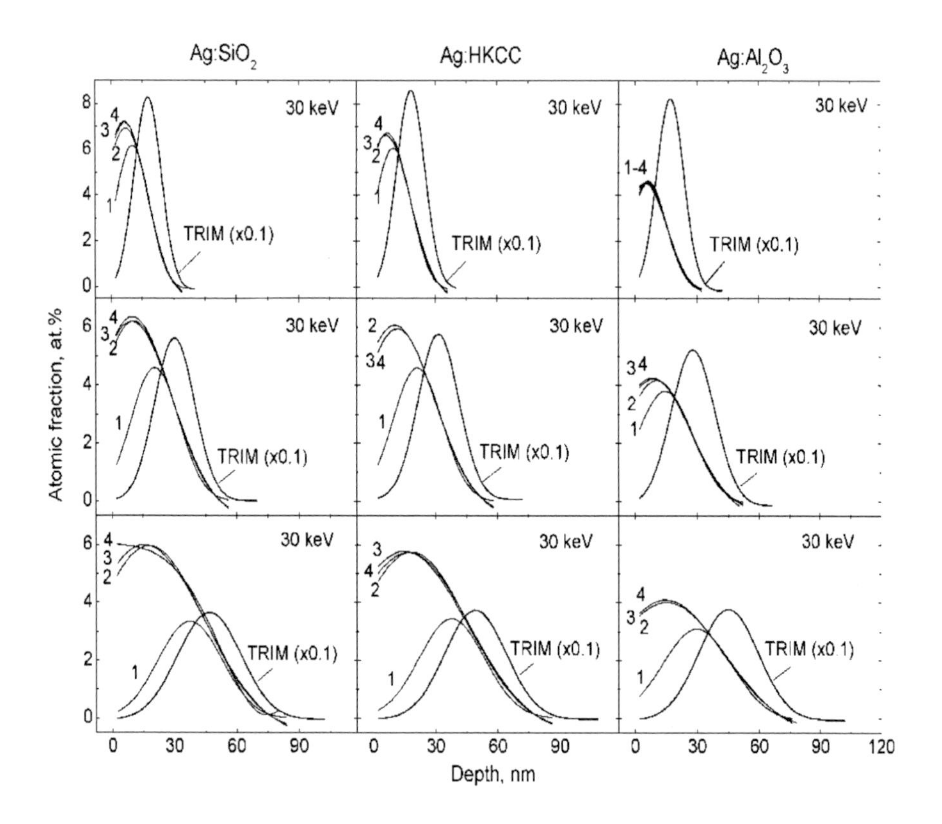

Figure 6. Calculated Ag-ion implanted depth distributions in amorphous insulators: SiO_2, Al_2O_3 and SLSG as a function of energy and dose: 1- 0.1·; 2- 0.3·; 3- 0.6· and 4- $1 \cdot 10^{16}$ ion/cm^2. There is also a profile corresponding to the TRIM simulations, which does not take into account sputtering and atom-target mixing effects [137].

All calculations were obtained at the dose simulations below 10^{16} ion/cm^2, because at higher dose implantation the increasing metal-ion concentration is above the solubility limit in these dielectrics [9]. This causes nucleation and growth of the MNPs that immediately alters the implanted ion penetration depth in the near-surface layer. Though it is impossible to calculate a correct DYNA ion-profile for high doses, nevertheless the metal distribution in implanted insulators for such cases may also be predicted from the present calculated data. Since both the increase of metal concentration in the depth profile and the sputtering yield depend on implantation time, then the metal particle nucleation and growth will also vary with time and depth. It is obvious that during implantation the size and growth of the particles with depth is *"proportional"* to the metal filling factor, because they are both determined by the ion concentration profile. Consequently, in accord with the calculated asymmetrical profiles for a dose of 10^{16} ion/cm^2 Fig. 6, the large MNPs (or/and the higher filling factor) in the same insulators implanted with higher doses will be close to the implanted surface, with small ones in the interior of the implant zone. These predicted features for implanted MNPs are qualitatively confirmed by the silver depth concentration in the SLSG derived from experimental Rutherford backscattering spectrometry (RBS) [133] corresponds to present calculations (Fig. 7).

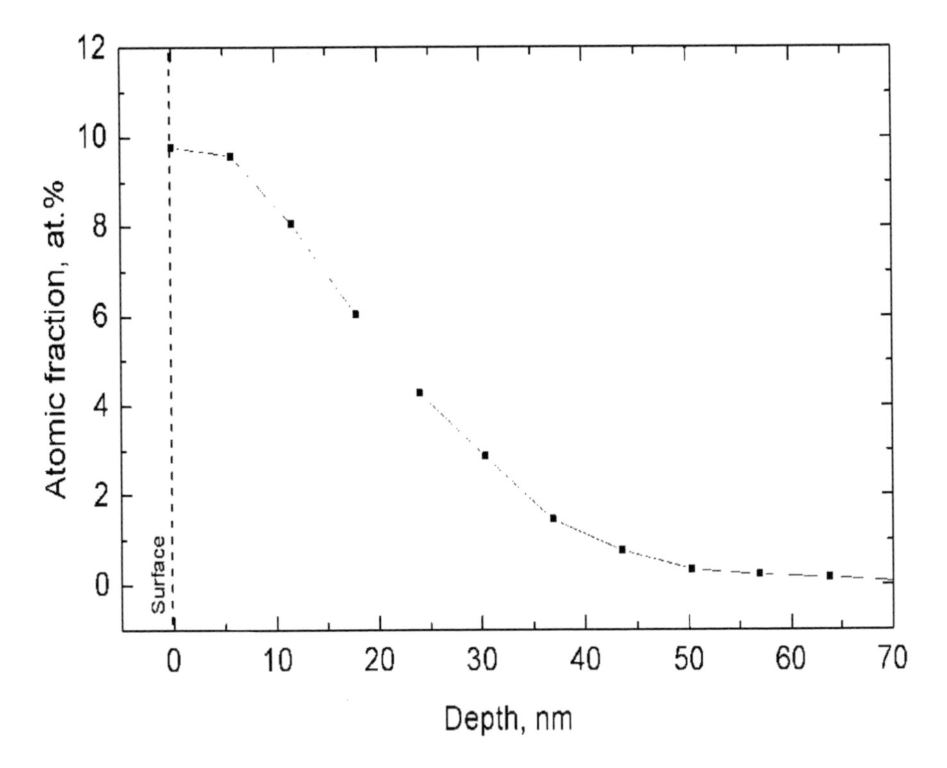

Figure 7. The depth distribution of silver derived from the RBS spectrum for ion dose of $7{\cdot}10^{16}$ ion/cm^2 at 60 keV into the SLSG [133].

Chapter 3

SYNTHESIS OF SILVER NANOPARTICLES BY LOW ENERGY ION IMPLANTATION

Ion implantation is an effective technological tool for introducing single impurities into the surface layer of the substrate to a depth of several micrometers. The degree of surface modification of the materials depends on their individual chemical and structural properties, as well as on variations of implantation parameters, such as the type and energy of an implant, current density in ion beam, substrate temperature, etc. A most critical parameter is ion dose F_0, which determines the implant amount. Depending on the modification of dielectrics by irradiation, ion implantation can be conventionally divided into low-dose and high-dose processes.

In the case of a low-dose irradiation ($\sim F_0 \leq 5.0 \cdot 10^{14}$ ion/cm^2), the Ag ions implanted after stopping and thermalization are dispersed throughout the volume of the dielectrics and are well separated from each other. The energy of the implant is transferred to the matrix via electron shell excitation (ionization) and nuclear collisions. This causes radiation-induced defects, which, in turn, may reversibly or irreversibly modify the material structure [9]. Various types of crystal structure damage have been observed in practice: extended and point defects, amorphization and local crystallization, precipitation of a new phase made up of host atoms or implanted ions, etc.

The range of high-dose implantation may be divided into two characteristic dose sub-ranges. In the range $10^{15} \leq \sim F_0 \leq 10^{17}$ ion/cm^2, the concentration of Ag ions exceeds the solubility limit of metal atoms in matrices and the system relaxes by nucleation and growth of MNPs (Fig. 8), as illustrated in plane [121] and cross-section [81] of transmission electron microscopy (TEM) views of SiO_2 glass with ion-synthesized Ag particles(Fig. 9 and 10). The threshold dose value (at which MNPs nucleate) depends on the type of the dielectric matrix and implant. For example, for 25-keV Ag+-ion implantation into $LiNbO_3$, the threshold dose was found to be $F_0 \sim 5.0 \cdot 10^{15}$ ion/cm^2 [33], for 30-keV silver ions embedded in epoxy resin, $F_0 \sim 10^{16}$ ion/cm^2 [170]. The next sub-range of high-dose implantation, $\sim F_0 \geq 10^{17}$ ion/cm^2, leads to the coalescence of already existing MNPs with the formation of either MNP aggregates or thin metallic quasi-continuous filmsnear the dielectric surface. For instance, the irradiation of silicone polymer-glass by 30-keV Ag ions at higher-than threshold-nucleation doses favors the formation of aggregate structures (Fig. 11) [171]. The MNP distribution established in the dielectrics after coalescence or Ostwald ripening may be dramatically disturbed by post-implantation thermal or laser annealing.

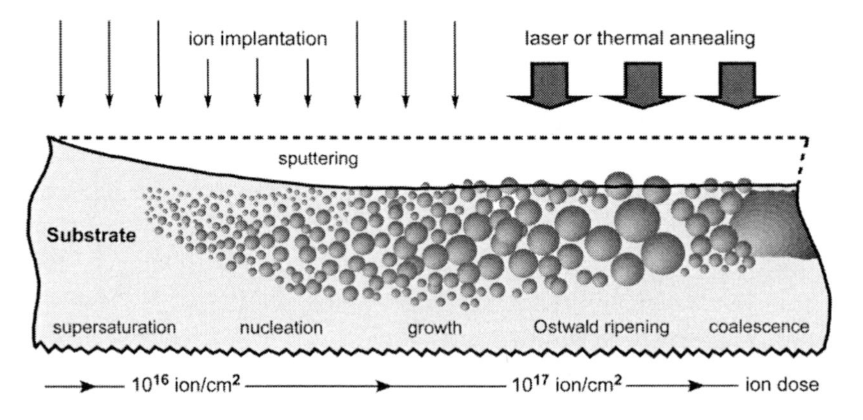

Figure 8. Basic physical processes (from left to right) involved in the formation of nanoparticle from an implant vs. the ion dose with regard to surface sputtering under irradiation.

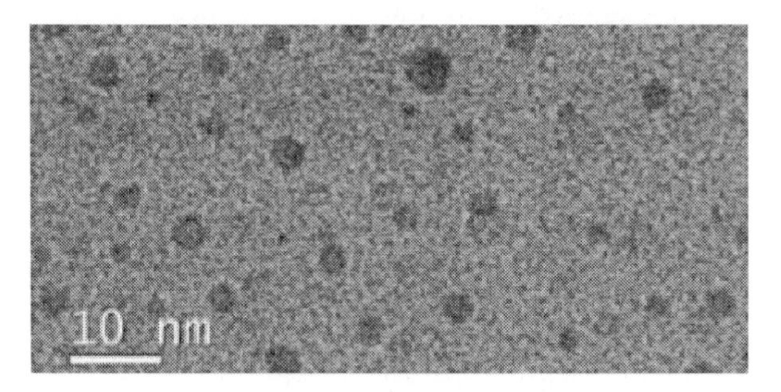

Figure 9. Plan-view TEM image of SiO_2 with Ag nanoparticles fabricated at a dose of $6.0 \cdot 10^{16}$ ion/cm^2 and an energy of 3 keV. Fragment of an image from [121].

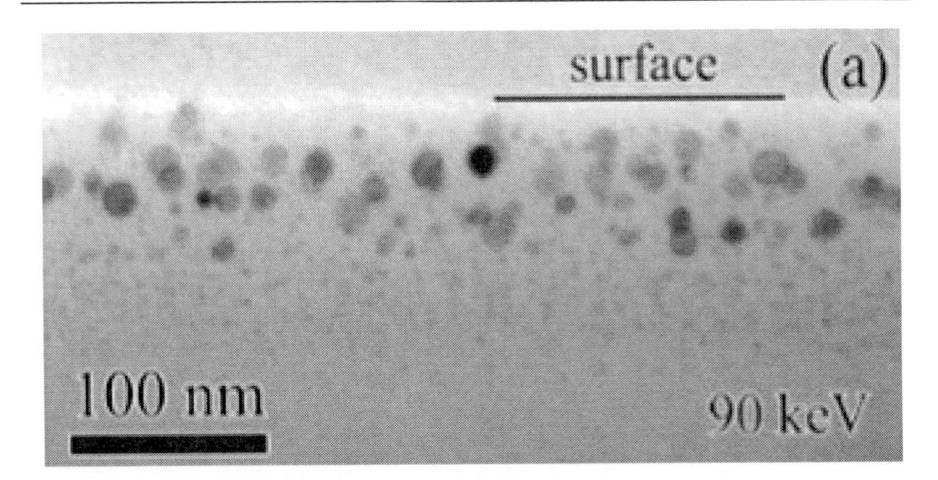

Figure 10. Cross-section TEM image of SiO_2 with Ag nanoparticles fabricated at a dose of $5.0 \cdot 10^{16}$ ion/cm^2 and an energy of 90 keV. Fragment of an image from [81].

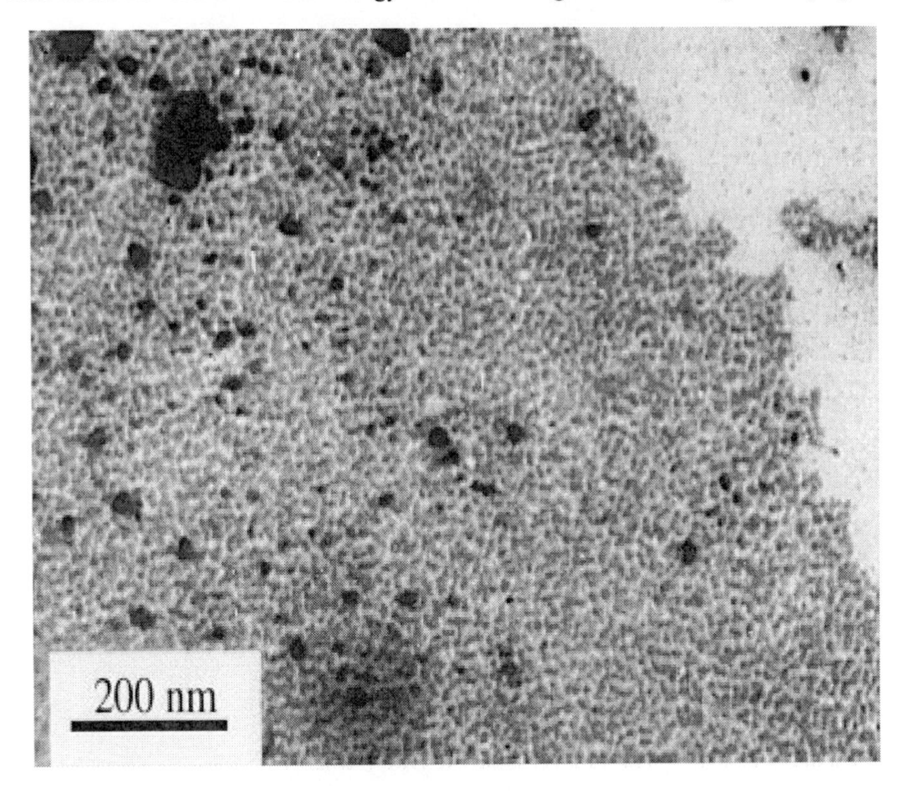

Figure 11.Plan-view TEM image of silicone polymer-glass with Ag nanoparticles fabricated at a dose of $3.0 \cdot 10^{16}$ ion/cm^2 and an energy of 30 keV [170].

Chapter 4

ION SYNTHESIS OF SILVER NANOPARTICLES IN GLASS AT ROOM TEMPERATURE

Although the implantation is made with Ag ions, the dynamics within the ion beam trajectory in the glass and the fact that there is a large capture cross-section for electronsof matrix atomsat low ion velocities, means that the Agion in dielectric will have a high probability of being in a neutral charge state (Ag^0) as it slows down. The mobility of the neutral atom is higher than that of the ion and additionally there are chemical reactions between the silver and the lattice ions. These are particularly difficult to assess in a target material such as SLSG as the surface chemistry of this multi-component glass is even more complicated than within the equilibrium conditions of the bulk material. Analyses of the surface show quite different depth distributions of the host elements, impurities and the tin dopants together with intrinsic structural defects (such as oxygen vacancy sites), and these as well as the dopant ions, exist in several valence states [172]. Within the glass medium there is competition between Ag and other ions for oxygen bond formation. However, the differences in Gibbs free energies can lead to Ag-Ag bond formation and hence aggregation of several Ag atoms. As was discussed [173], in spite of the fact that the free energy of silver oxide, at -2.68 kcal/mol at 25°C, is lower then that for pure metallic silver (0 kcal/mol at 25°C), the free energy of formation of SiO_2 (~ -200 kcal/mol at 25°C) is even lower. Consequently there is dissociation of Ag-O bonds to form Si-O and Ag-Ag bonds as this reduces the total energy of the system. These same arguments suggest that the silver nanoparticles have a sharp boundary between the silver and the host glass.

However, in some cases there is still evidence, which indicates that an outer silver oxide layer may act as an interface between the glass and the metal[56].

An excess of neutral Ag atoms in the glass, above the solubility limit, causes nucleation and growth of metal particles. If nucleation and particle growth result from attachment of neutral Ag atoms, then, if slow diffusion of substrate atoms is compared with the rate of incorporation of the implanted impurity species reaching the nucleation sites (diffusion limited growth), the attachment frequency is proportional to both the impurity diffusion coefficient and to the implant concentration [174]. Since the increase of Ag concentration in the depth profile depends on implantation time, the MNP nucleation will also vary with time and depth. In such a system the size of the growth particles with depth is partially determined by the ion concentration profile. As was shown above, for present condition of ion implantation the final Ag profile is characterized by a maximum concentration near the surface and differs from the theoretical symmetrical Gaussian distribution of the initial implantation. This implies the larger Ag nanoparticles are close to the implanted glass surface, with smaller particles in the interior of the implant zone. On the other hand, the concentration profile peak of implanted Ag ions moves during implantation, going deeper into the substrate as the sputtering, and hence, the nucleation and growth of metal particles is initiated at different depths, consistent with irradiation time and sputtering.

In practice, optical properties of Ag nanoparticles embedded into glasses are characterised by absorption and reflectance in the visible region. The intensity and spectral position of the peak depends on the concentration and size of the Ag particles, which in the case of spheres, are given by Mie theory predictions at longer wavelengths for large metal particles [175], and hence qualitative size estimates may be applied to the optical spectra. In Fig. 12 the reflectance of Ag-implanted SLGT corresponding different stages (different doses) of implantation at 60 keV and at various temperature of substrate are presented. At an early stage of implantation ($2 \cdot 10^{16}$ ion/cm^2) the smallest Ag particles appear in the glass at a depth consistent with the Gaussian distribution prediction. As seen in Fig. 12a, there is no remarkable difference in the reflectance peak positions (~ 450 nm) between samples prepared at various substrate temperatures. Increasing the ion dose leads to the appearance of reflectance peaks at different wavelengths and overall changes in the shape of the reflectance curves in dependence on temperature. For a dose of $3 \cdot 10^{16}$ ion/cm^2 the peaks shifts monotonically to a longer wavelength between 470 nm for a 20°C and 500 nm for the 60°C case (Fig. 12b). In these cases the

Ag particles become larger, though at higher temperature there are many much smaller MNPs then at 20°C.

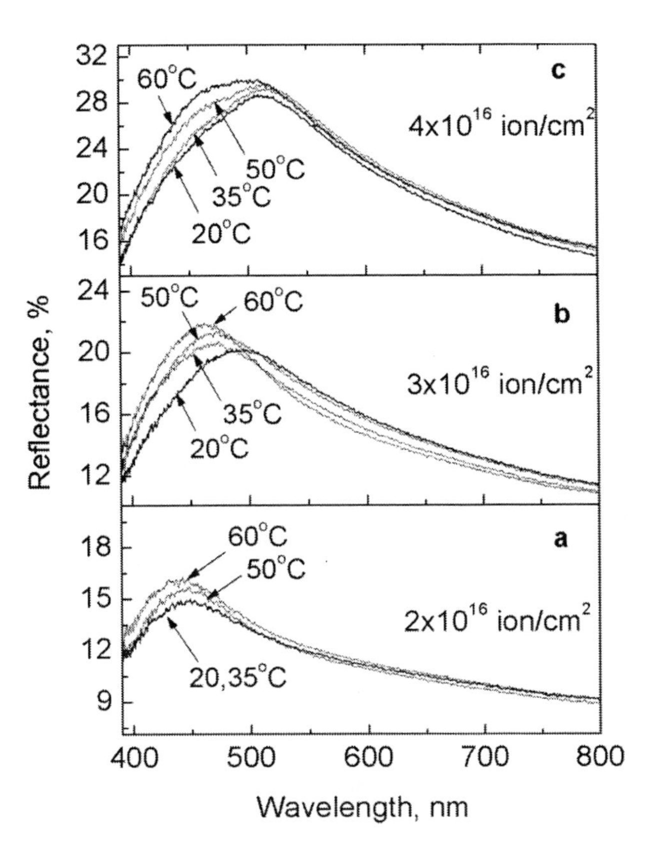

Figure 12. Reflectance of Ag-implanted SLSG at bulk-substrate temperatures of 20, 35, 50 and 60°C for various doses [146].

Most remarcable changes in reflectance spectra were registered at the highest dose of $4 \cdot 10^{16}$ ion/cm^2. The samples prepared at temperature higher then 35°C are characterised by reflectivity consisting of overlapping spectral bands with two maxima, for example at 470 and 510 nm for 50°C implants as shown by vertical arrows in Fig. 12c, and at least two distinct particle size ranges are favoured. However, in the cases of 20°C implantation, there is evidence of only one broad reflectance peak, also near 510 nm. These differences in reflectance spectra in Fig. 12c, and the corresponding models of the size of the Ag particles, cannot be described by variations in long-range Ag diffusion at 20 to 60°C (Fig. 2) only, though some differences in diffusion values for these temperatures, of course, essential. The measured RBS data for samples prepared at the highest temperature show that the width of the Ag depth penetration is approximately constant (Fig. 13) [134]. Also, the formation of MNPs at high dose appears in the glass layer over the same thickness range for the temperatures between 20 and 60°C. The explanation of the appearance of a bimodal concentration dependence, which has mainly large particles in the outer region and mostly small particles in the deeper zone, may result because of the variations of the Ag ion concentration into the glass. It was suggested for the case of higher energy (> 150 keV) Ag-implants into glass [32], that one depth region is set by the penetration maximum of the Gaussian concentration profile, and the second is at maximum of glass damage where there are peaks in the vacancy concentration, displaced atoms, pint defects and broken bonds. Similar consideration was also applied for the present case of low energy implantation. In Fig. 14 the concentration profile calculated from the TRIM programme, and the corresponding vacancy profile, are presented for the case of 60 keV Ag implants into SLSG using the SRIM-2000 programme [162]. It is seen that the maximum of glass-damage profile is resolved from the Ag concentration peak, and is placed closer to the irradiated glass surface. Taking into account the enhanced Ag damage-related diffusion to the surface, which is most effective at higher temperature (60°C), it is possible to explain the probabilities for accumulation of Ag atoms with the consequent growth of metal particles in the damage region. It should be noted that damage profiles move from the irradiated substrate surface, consistent with sputtering. Overall, the distributions of the impurity and damage profiles result in formation of bigger particles close to the glass surface, with a range of smaller particles deeper below the surface.

The reflectance spectra for similar samples Ag-implanted at the different temperature of SLSG from 60 to 180°C is presented in Fig. 15 [167]. The net reflectance and average particle size both decrease with higher temperature implants. The reflectance peak moves from 490 (60°C) to 450 nm (180°C) and the intensity decreases by ~14%. This optical result is consistent with RBS data, which shows that the high temperature implants lower the local concentration, both by inward diffusion and by enhanced sputter losses RBS spectra are plotted in Fig. 16 [167]. These indicate a sharp Ag peak at 60°C, but by 180°C, there is loss of Ag from the glass surface (~19 %) and Ag in-diffusion from the implanted layer.

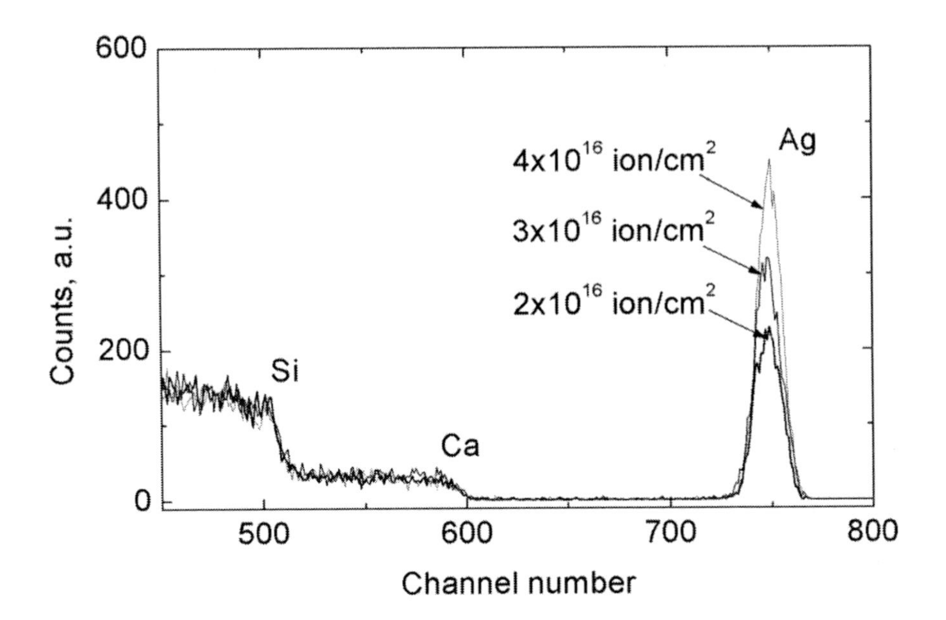

Figure 13. The RBS data for the Ag-implanted SLSG at a bulk-substrate temperature of 60°C for various doses [148].

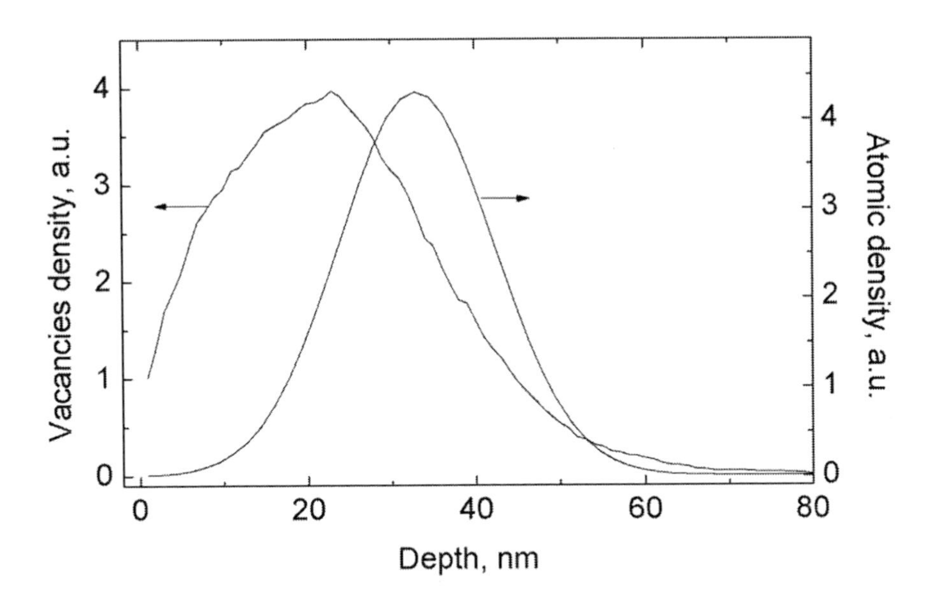

Figure 14. Calculated 60 keV implanted silver TRIM concentration profile in SLSG (right hand scale), and vacancy distributions (left hand scale), as characteristic of radiation glass damage without taking into account the sputtering process [210].

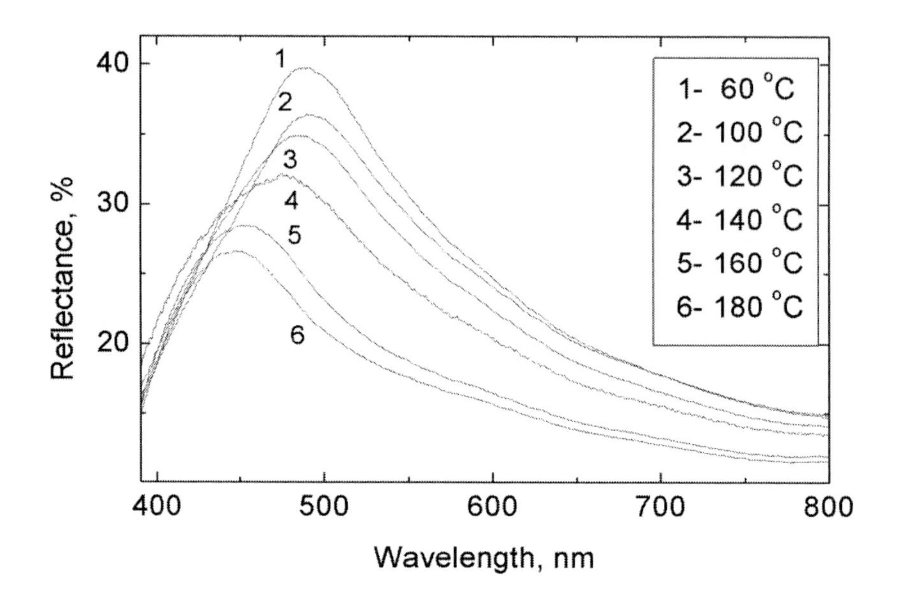

Figure 15. Reflectance of 60 keV Ag-implanted float glass for a dose of $4 \cdot 10^{16}$ ion/cm^2 at various SLSG temperatures [210].

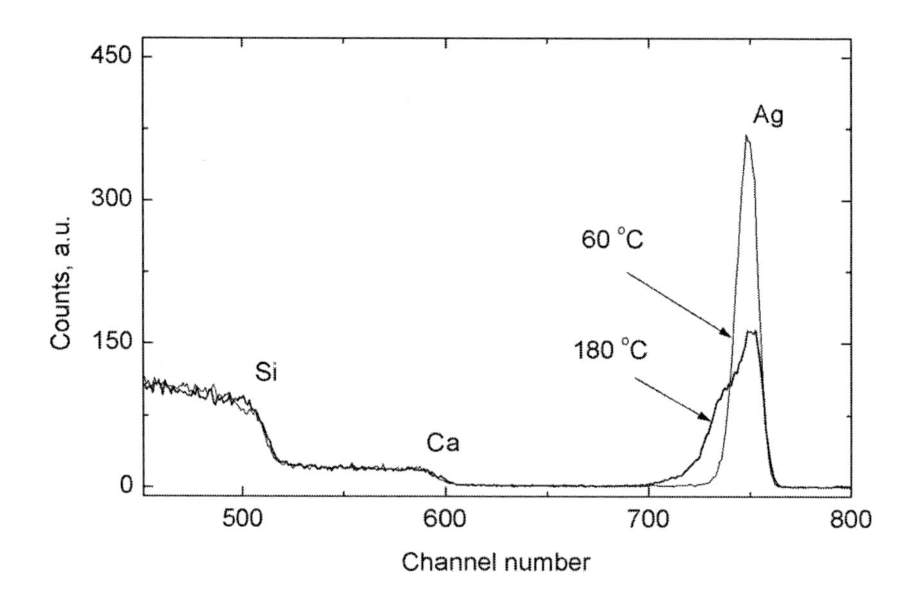

Figure 16. The RBS data for the 60 keV Ag-implanted float glass for a dose of $4 \cdot 10^{16}$ ion/cm^2 at surface-substrate temperatures of SLSG [167].

Thus the average Ag size decreases as seen by the reflectance data. Similar trends were exemplified in earlier RBS measurements [166] using implantation at the temperature from 250 to 600°C. For high-temperature Ag ion implantation into dielectrics, the diffusion coefficient drops dramatically after MNP formation.This means that the critical time for metal particle nucleation is the beginning of the implantation as it was suggested above from TRIM calculation for different temperatures of substrate, and therefore the substrate temperature during this initial phase. After particle nucleation has commenced, any changes, such as increasing temperature from beam heating or increases in ion current, will not interrupt the growth of metal particles. Conversely, high temperature conditions initially will increase the impurity diffusion and so reduce super-saturation and particle nucleation.

Chapter 5

OPTICAL REFLECTANCE OF ION-SYNTHESIZED SILVER NANOPARTICLES

Ion implantation gives the possibilities for the synthesis of MNPs in the volume of insulators with high values for the metal filling factor that lead to new perspectives for their opto-electronic applications. The optical properties of glasses containing implanted nanoparticles have been studied extensively by linear absorption spectroscopy, by the z-scan method or by direct measurements of the third-order optical susceptibility. The interpretation of experimental optical data is usually based on a restricted approximation in which the composite material acts as a dielectric medium containing equal-size MNPs, uniformly distributed in the total implanted volume. Moreover many authors assume that the absorption band is defined by measurements of transmission data only, which is incorrect [176]. To derive the absorption properties of a thin composite layer, one must separate effects of absorption from reflection in the measured transmission data. In a simplistic model of a uniform nanoparticle distribution throughout the bulk, there is no inconsistency, but this does not apply to the real ion-implanted material. As was discussed, one of the main features of the ion implantation process is a non-uniform statistical penetration of accelerated ions into the substrate that leads to the growth of metal particles with a very wide size distribution in the depth from the irradiated glass surface, as was shown by electron microscopy [57, 125]. Failure to include this non-uniformity causes considerable error in assessing the particle size distribution and in interpretation of optical properties. One of the possibilities for analysing of optical properties of dielectrics with non-uniform size distribution of refractive index over the depth is the consideration of the composite as consisting of a number of layers with MNPs with specific-sizes [176]. This approach could be also used for modelling and description of the optical reflectance of glass containing silver nanoparticles formed by ion implantation.

As was shown in Fig. 7, low energy implantation of metal ions leads to a non-uniform distribution in the glass, which is different from the statistical Gaussian profile. Silver depth concentration in SLSG derived from experimental RBS spectra shows a maximum near the implanted surface of the sample with some penetration to about 60 nm. The large silver nanoparticles in the glass are close to the implanted surface with small ones in the interior of the implant zone.

Optical spectra of such implanted glass are presented in Fig. 17. The transmittance spectrum is characterised by a deep minimum near 430 nm and the shape of spectral curve is almost symmetrical. The reflectance spectra are more complex and, although the transmission is the same whether the glass is viewed from the implanted or the reverse face, the shapes of the reflectivity curves differ. Overlapping peaks of reflectance spectra measured from the implant face of the samples exhibit a shoulder at about 430 nm, on the left side of a clearly determined maximum at 490 nm, whereas reflectivity from the rear face appears to have a simpler peak at longer wavelengths near 500 nm.

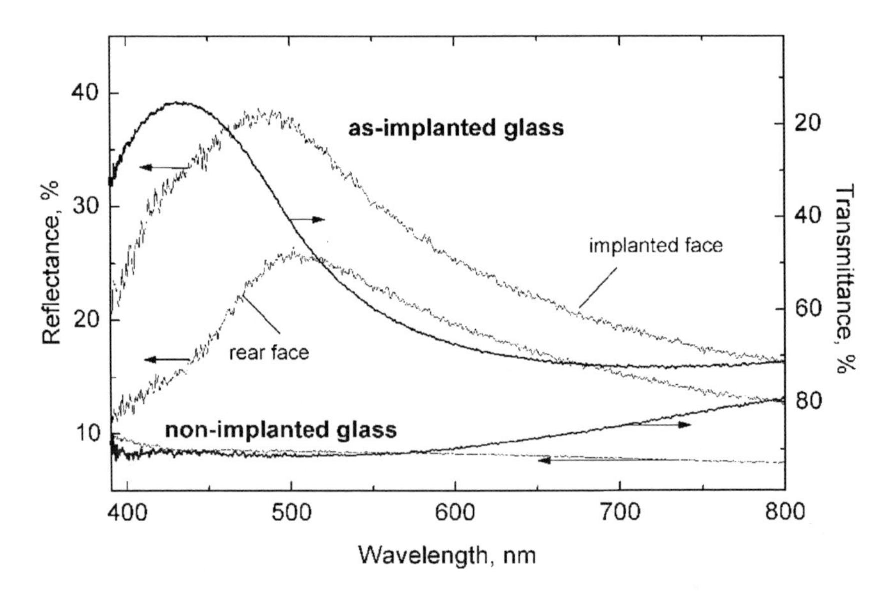

Figure 17.Optical transmittance and reflectance of the silver implanted SLSG and virgin glass with a dose of $7 \cdot 10^{16}$ ion/cm^2 and an energy of 60 keV.. Reflectance was measured from both the implanted and rear faces of the sample [132].

As the typical sizes of spherical metal particles formed by ion implantation are orders of magnitude smaller than the wavelengths of visible light [9] (Figs. 9 - 11), composite optical properties can be treated in terms of an effective medium theory. Also the laws of geometric optics for light beam directions and the Fresnel formulae for the intensities can be applied. An effective dielectric permeability, ε_{eff}, provided for dispersions of spherical metal particles with complex dielectric constants, ε_{Ag}, and a filling factor, f, in a surrounding glass (ε_m) for implanted composite may be derived from the effective medium theory, for example Maxwell-Garnet equation [177, 178]. For reflectivity evaluation of the composite material with a metal distribution which is changing in depth, and hence with a changing ε_{eff}, the implanted sample was considered as consisting of thin homogeneous isotropic layers characterised with their own constant ε_{eff} and f. For calculation of the multilayer reflectance, a matrix method [176] using the complex Fresnel coefficients was applied in this study for the case of normal incidence of the light. The values of f and thicknesses of composite layers may be estimated at different depths in the sample from Fig. 7. Assuming the surrounding glass to be a non-absorbing medium it should be noted that the refractive index of SLSG (1.54) increases after incorporation of dispersed silver ions in its volume [179].

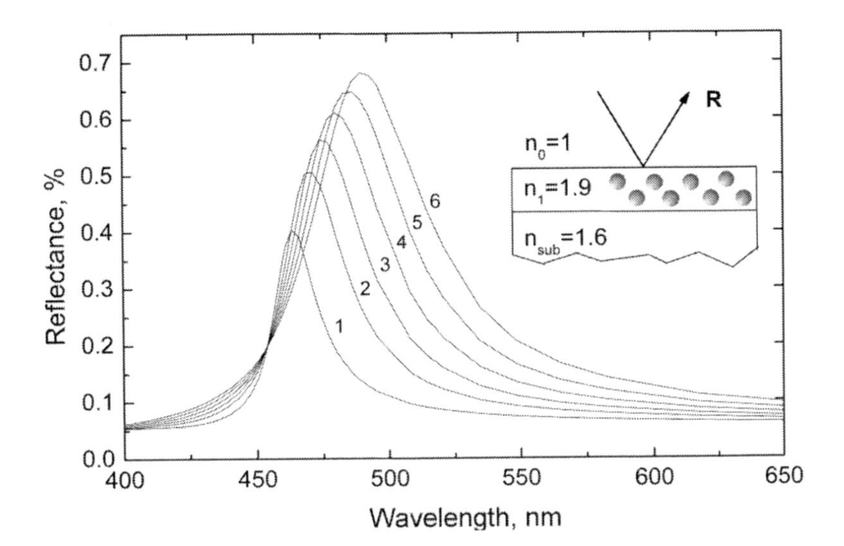

Figure 18. Calculated optical reflectance of the silver-glass composites. Spectral curves correspond to layers with a metal filling factor: (1) 0.05 , (2) 0.08 , (3) 0.1 , (4) 0.12 , (5) 0.14 , (6) 0.16 [136].

Firstly, consider the case of reflectivity from a single absorbing layer with ε_{eff} on a transparent substrate. Using the symbolic expressions derived from the matrix method for reflectance of such a structure [176], it is possible to calculate the optical spectra of the surface layer. For trial values of: a refractive index ($n_1 = 1.9$) of the layer containing MNPs, thickness of 9 nm, f from 0.05 to 0.16 and a refractive index of SLSG substrate with silver atoms as $n_{sub} = 1.6$, the set of computed spectra are presented in Fig. 18. The reflectance intensity increases and the position of the reflectivity peak shifts continuously toward longer wavelengths with increasing f, as expected for optical spectral bands when using the Maxwell-Garnet theory,as a single maximum corresponding to each f value. Hence, such a consideration cannot describe the experimental reflectance spectra with at least two overlapping peaks shown in Fig. 17, and modelling in terms of a single-layer structure with an average metal concentration(f)is not suitable. Proceeding to the next modelling case of two absorbing layers, each with a thickness of 9 nm on a transparent substrate generates data of the form shown in Fig. 19. Examples shown in Fig. 19a are modelled for the case of reflectance from the implanted face of an implanted sample where the top medium has a very low refractive index ($n_0 = 1$), a high index first metal layer with refractive index of $n_1 = 1.9$, and a second layer with $n_2 = 1.7$. The value of the substrate refractive index was the same as in the previous system at $n_{sub} = 1.6$. Optical spectra in Fig. 19a present the calculations when f in the surface layer is changed from 0.08 to 0.16, and the deeper layer has a constant $f = 0.05$. In Fig. 19a all the examples predict reflectivity from the implanted sample face to be characterised by two peaks: one at 430 nm and another between 440 and 480 nm. The second peak corresponds to the surface layer, where the increasing silver concentration (f) shifts the peak position towards longer wavelengths. However, there are clear differences between the experimental reflectivity spectra from the implanted and the rear faces of the layers (Fig. 17), and so calculated spectra for reflectivity from the substrate side of the same multilayer structure are presented in Fig. 19b. Again there are two wavelength peaks, at somewhat different positions in the spectra, and the more intense reflectance band corresponds to the deeper layer, (from this viewing direction). Although the layer with a low value of f is effectively the outer layer, for rear face reflectivity, the intensity of the reflectance band at 430 nm corresponding to this layer is weaker than the reflectance for the same layer when measured from the implanted face (Fig. 19a). The second essential feature is that the spectral peaks of the layers with high values of f appear at longer wavelengths for reflectance measured from the rear face. This calculation therefore predicts the pattern seen in the experimental reflectance data shown in Fig. 17, that emphasizes the differences between front and rear face reflectivity for non-uniform nanoparticle depth distributions and underlines the problem that simple analyses of the transmission and front face reflectivity data do not give all the information required to derive the optical absorption band shapes.

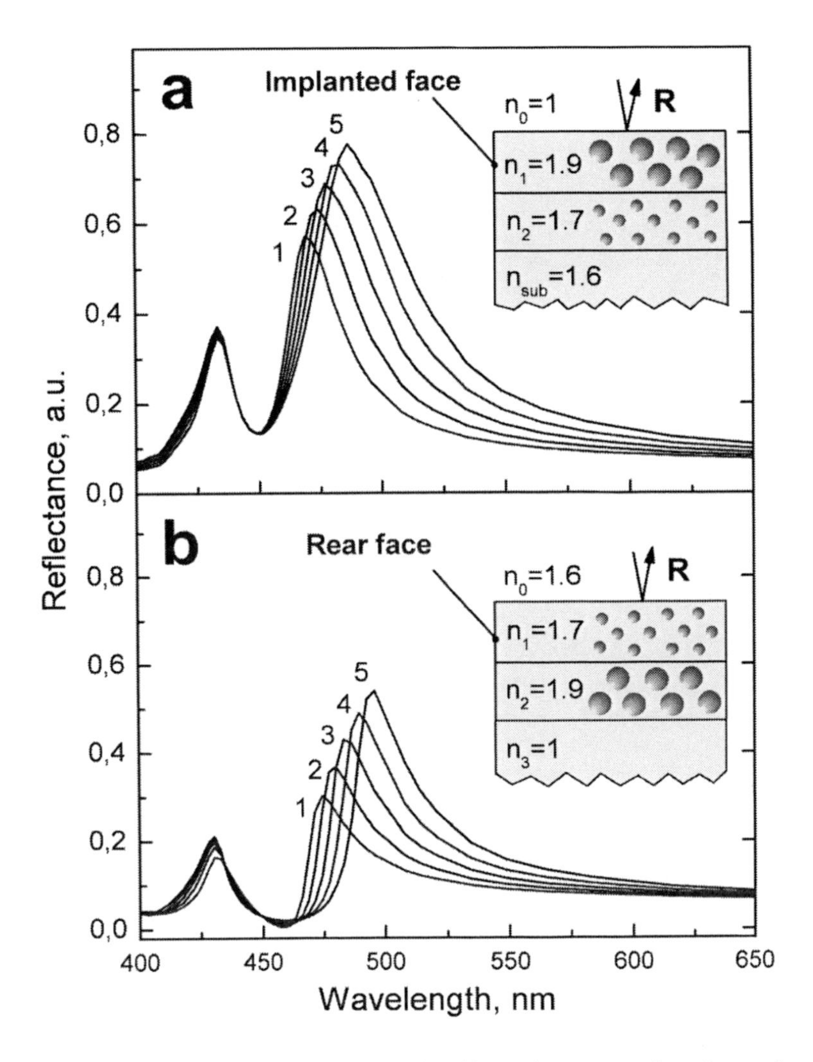

Figure 19. Calculated optical reflectance of the silver-glass composites. Spectral curves correspond to a layer a with refractive index of 1.9 and a metal filling factor: (1) 0.08 , (2) 0.1 , (3) 0.12 , (4) 0.14 , (5) 0.16. In the layer with refractive index of 1.7 the filling factor is 0.05. Figs. (a) and (b) correspond to implanted and rear face reflectivity, respectively [136].

Chapter 6

EFFECT OF SURFACE TEMPERATURE ON SILVER NANOPARTICLE FORMATION

Inevitably during ion implantation into dielectrics, which is a poor thermal conductor, there is a temperature gradient near the glass surface, as a nucleation and growth in the surface layer differ from those estimated by knowledge of the result of ion beam heating. Hence the effective temperature conditions for metal controlled substrate/holder temperature – bulk glass temperature. The measured parameter is therefore only a first step in the control process, although it can result in reproducible samples. To reveal the influence of the surface temperature gradient on the formation of metal nanoparticles in the glass implantation of thin (0.15 mm) and thick (3.1 mm) SLSG samples at the same temperature of 35°C were compared [142]. Both samples were fixed to water-cooled sample holed during implantation by thermoglue. It was assumed the surface temperature of the thick sample should be higher than in the thinner one and hence identical implant conditions will result in appearance of differences in the size of Ag nanoparticles, and their optical characteristics. Measurements of the transmittance and reflectance from both the implanted and rear face of the samples were made, and corresponding spectra are presented in Fig. 20. As seen from the figure there are no remarkable spectral differences between the transmittance curves from Ag nanoparticles into the samples near 425 nm, but some minor changes in the near-red transmittance. However, there are clear differences in the reflectance data. In previous paragraph the contrast between the information available from transmission and reflectivity has been stressed, and the changes are recognized as coming from the growth of Ag-implanted nanoparticles, which vary with depth into the glass surface.

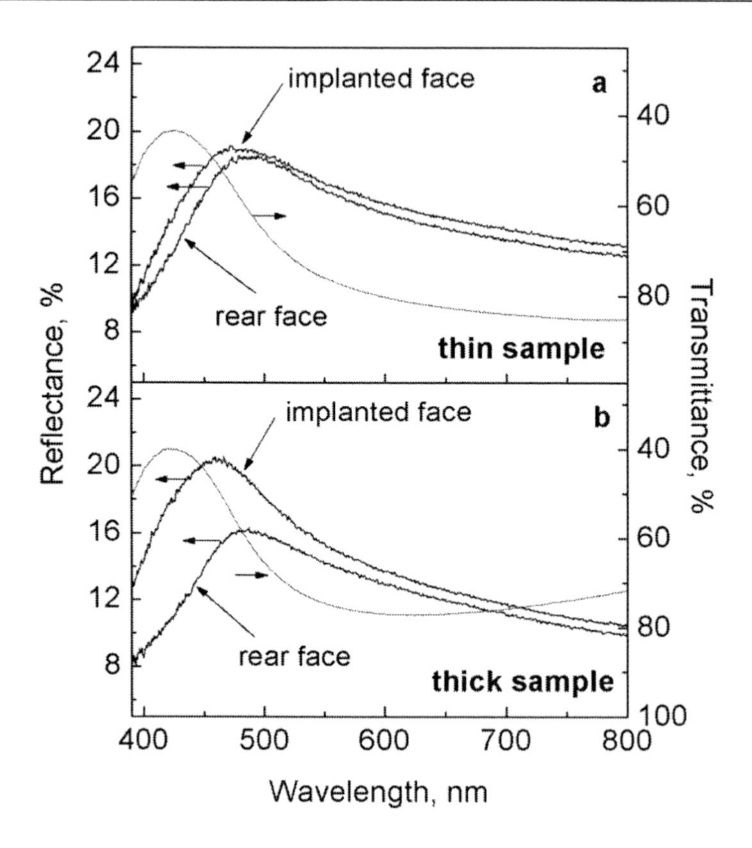

Figure 20. Comparison of the transmission (right hand scale) and reflectance (left hand scale) of 60 keV Ag-implanted SLSG with dose of $3\cdot10^{16}$ ion/cm^2 and at bulk-substrate (target holder) temperatures of 35°C for various thicknesses of irradiated substrate: a.- this sample of 0.15 mm; b.- thick sample of 3.1 mm. All the samples were measured from the implanted and rear faces [142].

The differences between implanted and rear face reflectivity of a thick sample, which contain peaks near 450 and 475 nm, (Fig. 20b), immediately emphasise that the distribution of particle sizes vary with depth beneath the implant surface. Although the transmission is the same whether the glass is viewed from the implanted or the reverse face, the shapes of the reflectivity curves differ. Whilst the reflectance differences from implanted and rear faces are monitoring an asymmetry of the nanoparticle size distribution and concentration of the particles with depth into the sample, the precise distributions cannot be determined. As already mentioned, particles of larger size are concentrated near the implanted glass surface, whereas small ones occur throughout the ion range. The reflectances from thin and thick samples are very different in spectral shape, although the RBS data for samples in both cases show an approximately constant width for the Ag distribution profile. Moreover, the reflectance from implanted and rear faces of thin samples, with peaks near 470 and 480 nm, are very similar in shape and intensity to each other (Fig. 20a). This suggests that the smaller temperature gradient across the glass results in a more symmetrical particle size distribution with depth. Moreover, the position of reflectance peaks of the thin sample is at a longer wavelength than in thick samples, indicating formation of a more uniform distribution of large particles.

Quite clearly the reason for differences in reflectance between thick and thin samples results from the different temperature gradient at the irradiated surface, and as seen from data in Fig. 20 suggests that for thin samples there is closer control to the base temperature of 35°C. For thin samples a more uniform particle size-depth distribution was produced than for thick glass targets. Since such temperature gradients and average temperature differences exist relatively close to room temperature, it is worth noting that this is contrary to some of the early models for describing the nanoparticle formation in insulators by ion implantation which are based on thermal spike considerations, as these assume the local temperature inside an ion trajectory within a silicate glass to be ~ 3,000 K [180]. Such mechanisms would not respond in the way described here. Modeling suggests that radiation damage enhanced Ag diffusion in glass is important, as are the temperature gradients, which develop in the surface of the insulator during implantation. Overall the initial beam and temperature conditions have a major influence on the resulting nanoparticle generation.

Chapter 7

SILVER NANOPARTICLE FORMATION AT DIFFERENT CURRENT DENSITIES

The fabrication of silver nanoparticles in a dielectric matrix by ion implantation is a complex process which depends on a number of factors. The conditions of metal nanoparticle synthesis can be varied depending on the ion implantation parameters such as ion energy, dose, ion current, target temperature, etc. In previous paragraphs it was that temperature of the irradiated glass is a significant factor for size control of the MNPs. Unfortunately, the target temperature is often ignored in experiments. Let us now to consider an influence of the ion current density and concomitant thermal effects on the silver nanoparticle formation and surface modification under the low-energy ion implantation of silicate glass (SiO_2).

Formation of the Ag nanoparticles in the implanted SiO_2 was estimated by the optical transmittance measurements showing an appearance of the characteristic band of SPR absorption. It was observed[151] that the increase of the ion current density monotonically shifts the absorption band to longer wavelengths indicating the rise of the nanoparticle sizes (Fig. 21).

AFM image of the virgin glass surface, which is relatively smooth, is shown in Fig. 22. AFM images in Fig. 23 show the glass surface morphology produced by the Ag^+ ion implantation at different current densities in the beam[151]. Compared to Fig. 22, formation of semi-spherical hillocks is observed for all implanted samples. This surface structure is explained by the sputtering of glass layer resulted in partial towering of the spherical-shaped metal nanoparticles nucleated in the near-surface layer of the substrate. Similar morphology was earlier detected by AFM for different metal nanoparticles synthesized in various dielectrics by low-energy implantation, for example: Ag ions into Ta_2O_5, SiO_2, Si_3N_4 [21, 66]. It is seen from the images of Fig. 23 that the hillock size (or particle sizes) increases with the ion current density. This result is in good correlation with the observed optical dependences (Fig. 21).

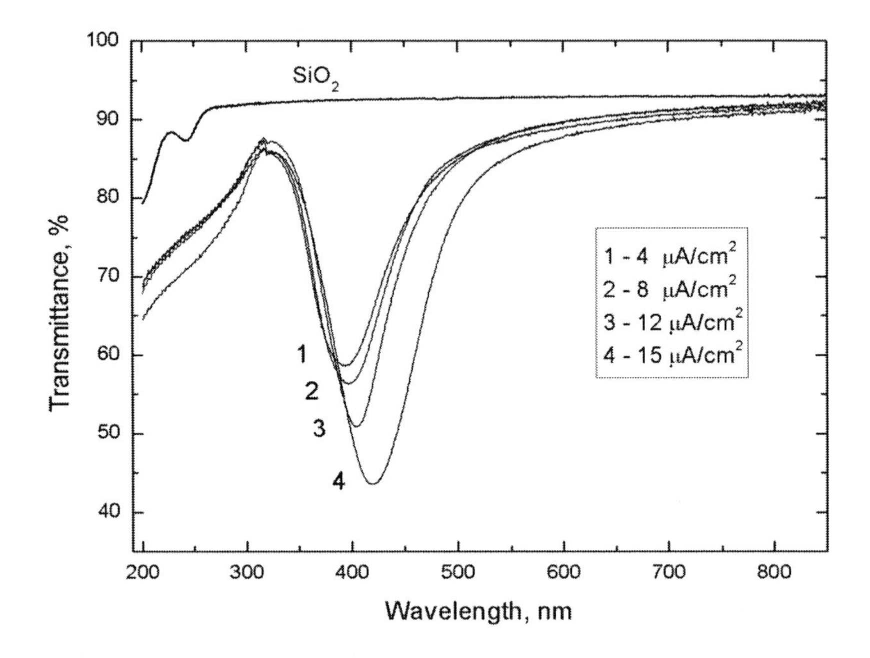

Figure 21. Transmittance spectra of SiO_2 samples after Ag^+ ion implantation with dose of $5 \cdot 10^{16}$ ion/cm^2 at various ion current densities [151].

The formation of bigger particles at higher ion current densities, when the dose is constant, may be explained by an increase in Ag atom mobility and faster particle nucleation. The increase in the diffusion mobility is expected due to the substrate heating by the implantation at high dose rate. The numerical estimation presented in shown that the coefficient of diffusion of silver atoms in the glass increased for two orders of magnitude with the substrate temperature rise from 20 to 100°C (Fig. 3). At the beginning of implantation, all samples in our case were at the same room temperature but, it is obvious, that by the moment of collection of the ion dose the substrate implanted at higher ion current density has higher temperature. Thus, the change in ion current density under implantation of metal ions into dielectric considerably affects the formation of metal nanoparticles. This method can be used for control of particle size to synthesize the metal/dielectric composites with desirable parameters.

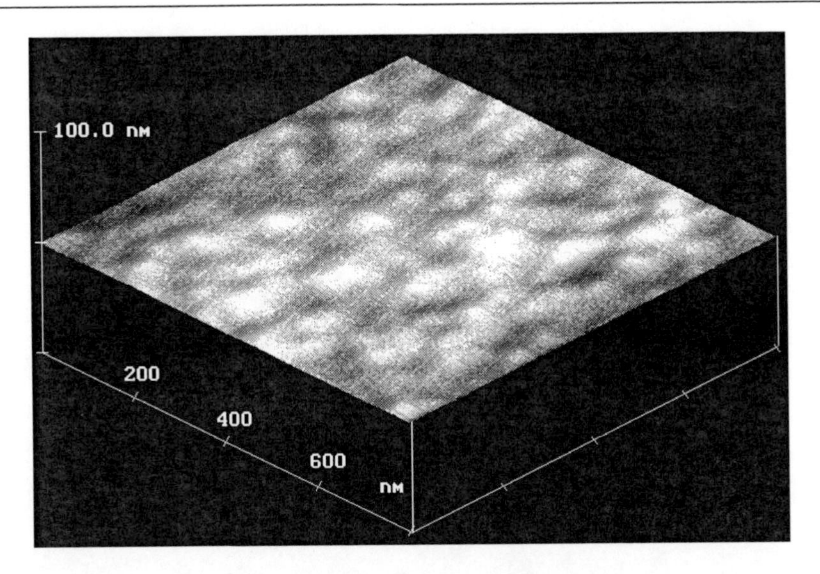

Figure 22. AFM image of SiO_2 surface before ion implantation [151].

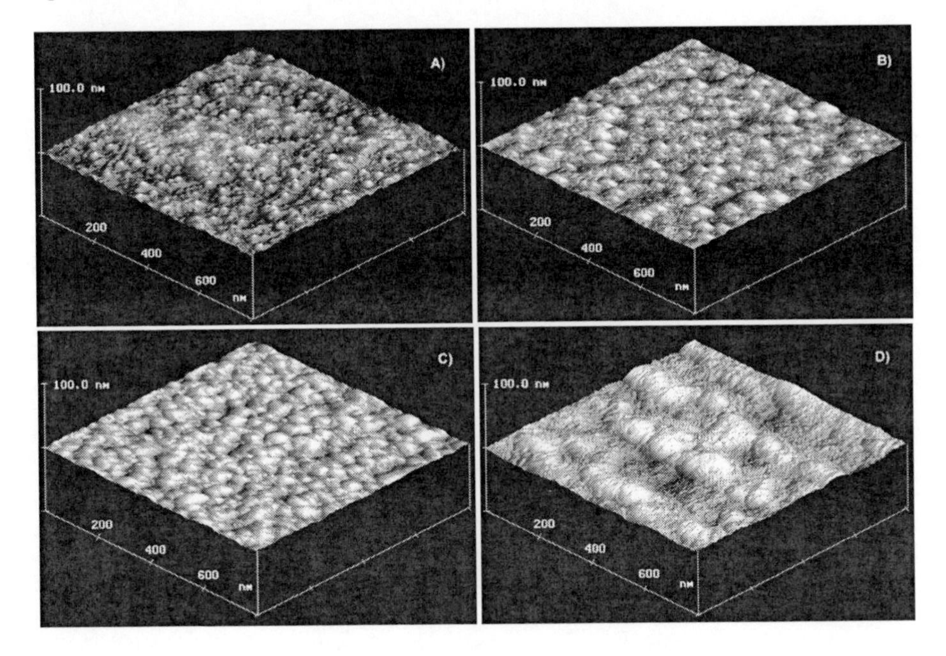

Figure 23. AFM images of SiO_2 surfaces implanted at different ion current densities: a) 4; b) 8; c) 12; d) 15 $\mu A/cm^2$ [151].

Chapter 8

THERMAL ANNEALING OF DIELECTRICS WITH IMPLANTED SILVER NANOPARTICLES

Ion implantation and diffusion of metals into insulators can lead to an excess of the metal which is unstable in the form of atomically dispersed particles. The systems relax into precipitates of metal nanoparticles. As was considered, one problem is that, as formed, there is invariably a wide distribution in the sample depth in terms of nanoparticle size. Many features, such as the optical response, are size-dependent; hence, for any potential application, variations in size degrade the performance, or confuse the interpretation of the fundamental processes. Consider the possibility of the furnace annealing for modification of metal nanoparticles in implanted dielectrics. Again, as in previous case, the size distribution of metal particles is monitored by optical transmittance and reflectance from both the implant and rear face of the samples. The difference between the data indicates the asymmetry in the size distribution with depth.

For the Ag ion implantation of SLSG at 60 keV and the dose of $7 \cdot 10^{16}$ ion/cm^2 is characterised by optical transmission, maximised near 430 nm, with a long wavelength tail extending across the entire visible range(Fig. 24a) [135]. Reflectance spectra measured from the implant face of such samples partially resolve overlapping peals, as a clearly determined maximum at 490 nm and the 430 nm shoulder, where reflectivity from the rear face is peaked more obviously at 500 nm. Note that there are major differences shown in Fig. 24a between assessments of the peak optical changes, and widths, for transmission, and front and rear face reflectivity.

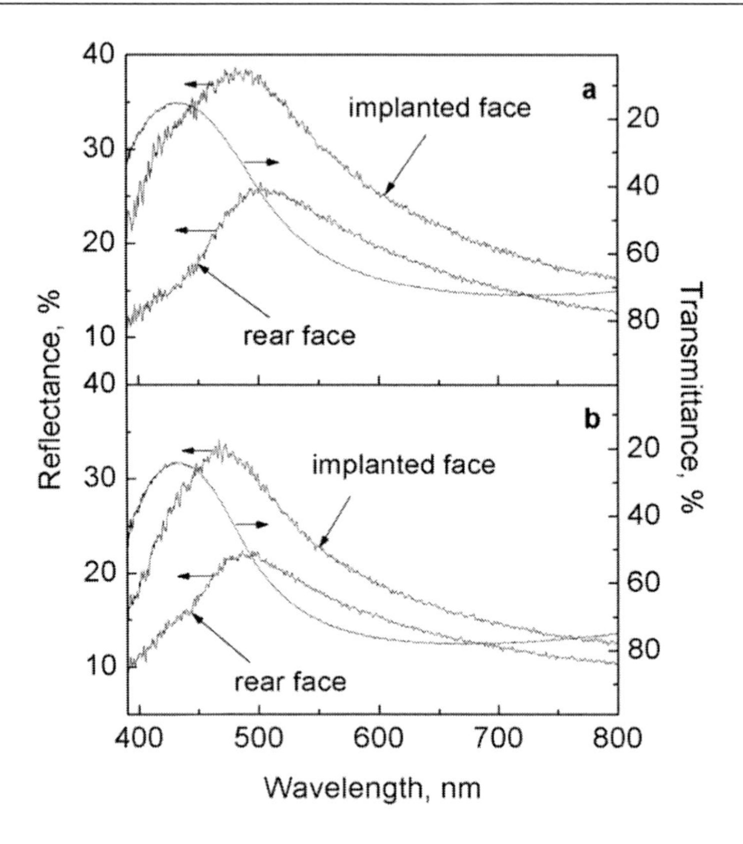

Figure 24. Comparisons of the transmission and reflectivity spectra of ion implanted silver in SLSG: (a) for the sample after implantation; (b) the spectra after a furnace annealing at a temperature of 350°C for 3 h in atmosphere [135].

The simplest method of altering the structure and distribution of the metal nanoparticles is to use a furnace annealing. The treatment can allow a mixture of diffusion, dissolution of the large clusters, or growth from small particles. These competing factors are strongly temperature dependent, but for the present purposes, the example from only a single temperature heating, of 350°C, are presented in Fig. 24b[135]. For studies of nanoparticles, it is essential to note that the results can be influenced by the rate of cooling from the high temperature. The magnitude of the signals, and their wavelength maxima, are markedly different for front and rear face reflectance measurements as in the case of as-implanted samples. In another words, there is the contrast between reflection spectra, which demonstrates the wide size distribution of nanoparticles in the depth. Ideally, a narrow size distribution would introduce the symmetrical reflectance bands.

Chapter 9

NONLINEAR OPTICAL ABSORPTION OF ION-SYNTHESIZED SILVER NANOPARTICLES

The Ag nanoparticles doped in different dielectrics demonstrate variable nonlinear optical properties [181]. The interest on such structures is based on the prospects of the elaboration of optical switchers with ultrafast response, optical limiters, intracavity elements for mode locking and sensors. Ag nanoparticles have an advantage over other metal nanoparticles (i.e., gold and copper) from the point of view that the surface plasmon resonance energy of Ag is far from the interband transition energy. So, in the silver nanoparticle system it is possible to investigate the nonlinear optical processes caused solely by SPP contribution.

Previous studies of nonlinear optical parameters of silver nanoparticles-doped glasses were mostly focused on determination of third-order nonlinear susceptibility$\chi^{(3)}$.It was predicted that silver-doped glasses possess by saturated absorption. The spectral dispersion of the imaginary part of third-order susceptibility $\mathrm{Im}\chi^{(3)}$ of silver-doped glass matrices was analyzed and it was shown that $\mathrm{Im}\chi^{(3)}$ was negative in the spectral range of $385-436$ nm [182]. The nonlinear absorption coefficient β is also negative in the case of saturated absorption. The saturated absorption in silicate glasses with ion-synthesused Ag nanoparticles at wavelength of 532 nm and their dependence on laser radiation intensity are considered at present paragraph.

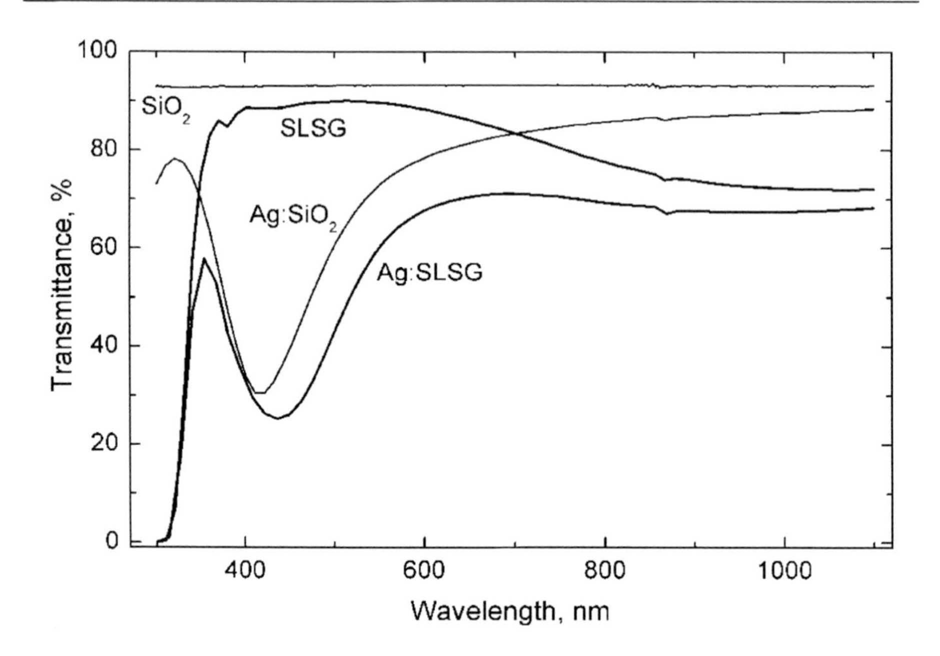

Figure 25. Optical transmittance of the Ag nanoparticles formed in SLSG and SiO_2 by implantation with a dose of $4{\cdot}10^{16}$ ion/cm^2 and an energy of 60 keV [183].

The Ag nanoparticles ion-synthesized in SLSG (Ag:SLSG) and (Ag:SiO_2) demonstrate the SPP band with minimum transmission in the range of 410–440 nm (Fig. 25) [183]. The normalized transmittance dependences of Ag:SLSG and Ag:SiO_2 samples measured using open aperture Z-scan scheme at laser radiation intensity of $I_0 = 2.5{\cdot}10^9$ W/cm^{2}and pulse duration of 55 ps is presented in Fig. 26[184]. The transmission of samples was increased due to saturated absorption as they approached close to the focal plane. After fitting of experimental data the β are $-6.7{\cdot}10^{-5}$cm/W in Ag:SLSG and $-3.6{\cdot}10^{-5}$ cm/W in Ag:SiO_2. Coefficient β can be presented as $\beta = \alpha/I_s$ where is I_s saturated intensity. The values of $I_s$are $1.1{\cdot}10^9$and $1.4{\cdot}10^9$W/cm^{-2}, also the Im$\chi^{(3)}$ are -$2.4{\cdot}10^{-8}$and $-1.3{\cdot}10^{-8}$esu in Ag:SLSG and Ag:SiO_2, respectively. In Figs. 27 and 27 values of β in dependence of laser intensity varied from $10^9$to $2{\cdot}10^{10}$ W/cm^{2}are presented [184]. As seen from the figures there are a decrease $\beta$of for higher intensities. In particularly, a 21- and 12-fold decrease of β was measured at $I_0 = 1.15{\cdot}10^{10}$ W/cm^{2}for Ag:SLSG and Ag:SG, respectively, compared to β detected at $I_0 = 1{\cdot}10^9$ W/cm^2.

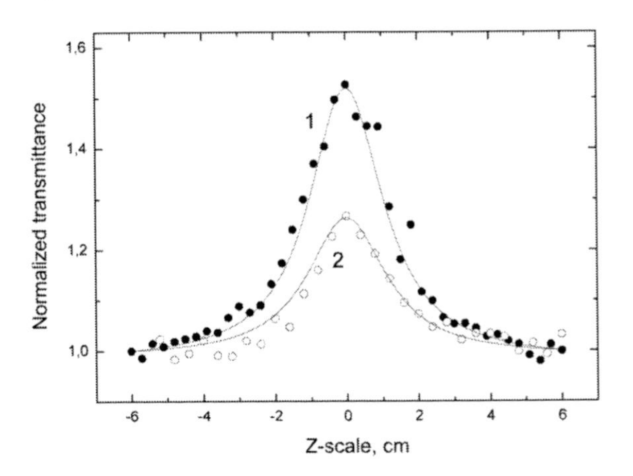

Figure 26. Normalised transmittance Ag:SLSG (1) and Ag:SiO_2 (2) samples at laser radiation intensity of $I_0 = 2.5 \cdot 10^9$ W/cm^2. Solis lines show theoretical fittings [184].

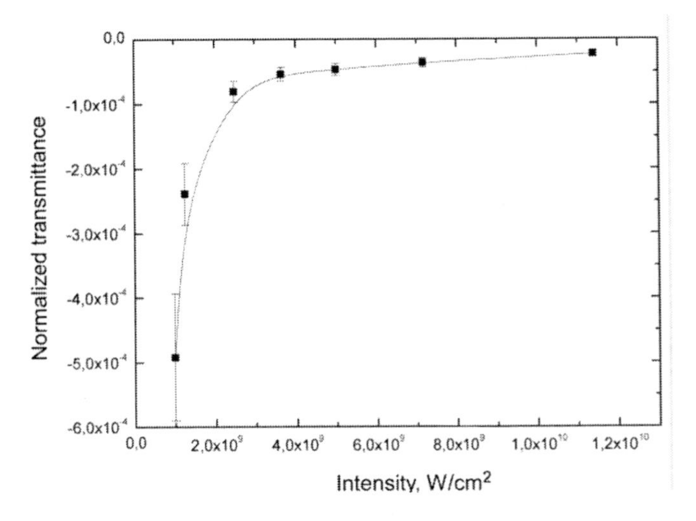

Figure 27. Coefficient β of Ag:SLSG in dependence of laser intensity [184].

The variations of transmission in similar MNP structures were attributed in some cases to the fragmentation [185-187], or fusion [188] of nanoparticles following the photothermal melting. It was reported about the alteration of the sign of nonlinear refractive index of small Ag clusters embedded in SLSG [189]. They noted that thermal effects could change the properties of nanoclusters. The transparency in these samples was associated with oxidation of Ag nanoparticles. However, no irreversible changes of transmittance were observed in present experiments.

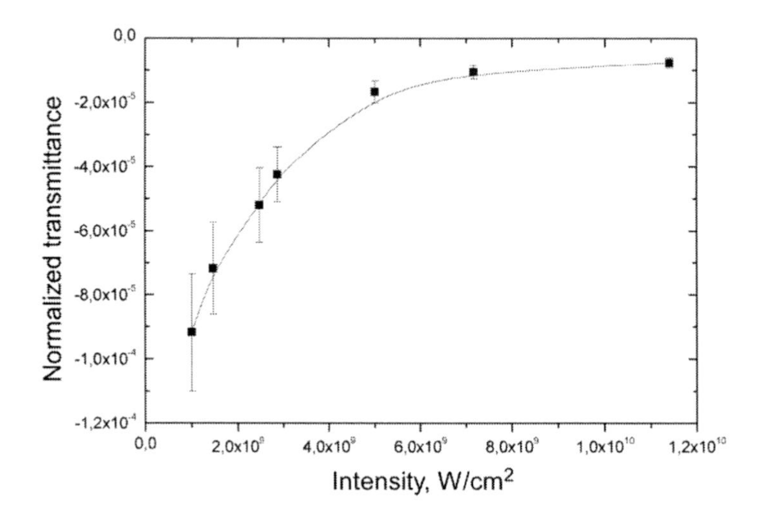

Figure 28. Coefficient β of Ag:SiO_2 in dependence of laser intensity [184].

The reverse saturated absorption can be responsible for the decrease of negative nonlinear absorption of Ag nanoparticles and it could be assume that in the case of picosecond pulses the reverse saturated absorption starting to play an important role in the overall dynamics of nonlinear optical transmittance of metal nanoparticles contained compounds, taking into account the saturation of intermediate transitions responsible for saturated absorption. Thus, saturated absorption in Ag:SLSG and Ag:SiO_2 was dominated at small intensities and decreased with the growth of intensity due to influence of competing effects, whereas the self-defocusing at low intensities was changed to self-focusing at high intensities. The possible mechanism of the decrease of Im$\chi^{(3)}$ is the influence of nonlinear optical processes with opposite dependences on laser intensity, also such as two-photon absorption [181]. The wavelength range corresponded to the interband transitions in Ag is located below 320 nm, so the two-photon absorption connected with interband transitions can be involved in the case of 532 nm radiation. The possibility of two-photon absorption due to interband transition of photoexcited electronswas previously demonstrated for Ag particles [58]. The three-photon absorption connected with interband transition for Ag nanoparticles was analyzed in [190]. Thus, saturated absorption in Ag:SLSG and Ag:SiO_2 was dominated at small intensities and decreased with the growth of intensity due to influence of competing effects, whereas the self-defocusing at low intensities was changed to self-focusing at high intensities.

Chapter 10

LASER ANNEALING OF GLASSES WITH ION-SYNTHESIZED SILVER NANOPARTICLES

Despite advantages the use of ion implantation for nanoparticle synthesis there have not yet emerged clear mechanisms which allow precisely controlled particles sizes and depth distributions. Latter has a certain drawback, which is the statistically non-uniform depth of penetration of implanted ions into a material. As was shown in previous paragraphs this leads to a wide size distribution of synthesized nanoparticles not only in the plane parallel to the irradiated surface but to a great extent also over the depth of the sample. Dispersion of nanoparticles with respect to sizes leads to a broadening of the SPR optical absorption band accompanied by a decrease it in the intensity [4]. This is also attributable to the dependence of the SPR spectral position on the particle size, i.e. the absorption spectrum in real sample is a superposition of several overlapping less intense bands that corresponding to particles of various sizes. The concern of the modern task is to increase the uniformity of the size distribution of MNPs synthesized by ion implantation using an approach of high-power pulse laser annealing with sequential furnace one. Experience gained from using the laser annealing techniques for various purposes allowed MNPs to be modified in various dielectrics. The main feature most of all the experiments with laser annealing of composites with MNPs is that the laser light was applied directly into the spectral region of the transparency of the dielectric matrix, and consequently, the intense laser pulses were primarily absorbed by the metal particles. Contrary to that, a new approach for annealing was demonstrated, when soda-lime silicate glass (SLSG) with Ag particles was irradiated by a laser light at wavelengths of glass absorption in the ultraviolet region [126]. When applying high-power excimer ArF (193 nm) laser pulses, a decrease of the reflectance intensity of composite samples was observed. It was suggested that the implanted silver particles in glass can be dissolved and the glass matrix can be modified to be a silver rich metastable new glass phase. If this is correct then the new phase will be the potential to be destabilized to precipitate out the new silver particles in a controlled fashion by furnace.

As was shown in this chapter the large ion-synthesized silver in the glass are close to the implanted surface with small ones in the interior of the implant zone. These features can be recognized in optical spectra of dielectrics with implanted nanoparticles (Fig. 17) [132]. The atomic force microscope images implanted surfaces of this sample shown in Fig. 29 [132]. As seen from the figure the implanted surface is smoother (roughness) and there are many hemispherical hills on this surface with an average diameter of approximately 100-150 nm. There are no such protrusions on the un-implanted sample. The reason for the existence of surface hills is assumed to be from a sputtering of irradiated glass during implantation, which leads to unequal ejection of ions of different elements from the surface, exposing the synthesized nanoparticles in the sub-surface glass. The sputtered glass thickness is typically of the order of tens of nanometers for the present ion dose [167], and, hence, the synthesized buried MNPs appear near to the glass surface in the implanted sample.

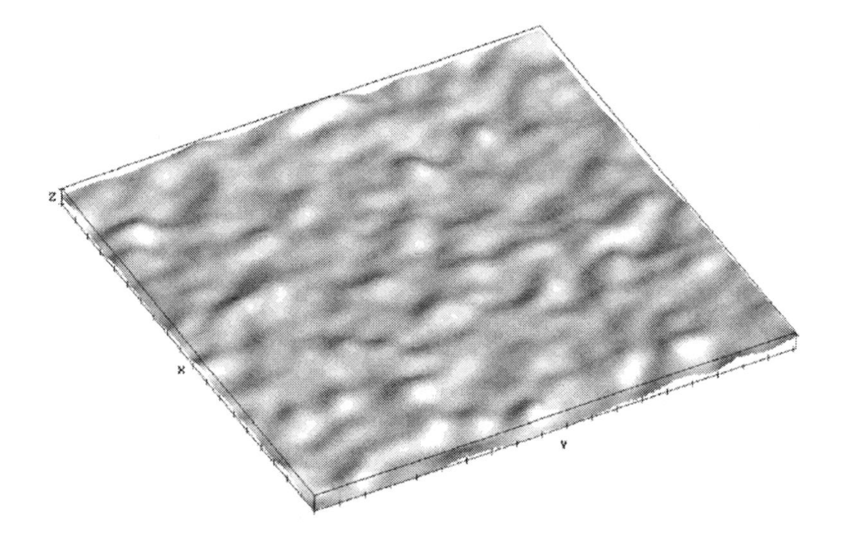

Figure 29. An AFM image as a top view under lateral illumination of the surface of SLSG Ag-implanted with a dose of $7 \cdot 10^{16}$ ion/cm^2 and an energy of 60 keV. The step along the X and Y axes id 100 nm, and the step along the Z axis is 3 nm [132].

Consequences of the excimer laser pulse with nanosecond pulse width and high beam intensity are heating, melting and/or vaporization (ablation) of material on a time scale of nanoseconds to microseconds. The excimer laser treatment has been applied to many glasses, but there is less information on high-power pulse laser interaction with dielectrics containing metal nanoparticles. In the present case the energy density is lower than the value of the ablation threshold for the SLSG, which was determined to be about 5 J/cm^2 [191]. Also, the excimer laser is characterized by a UV-wavelength, which is much longer than the typical sizes of the nanoparticles formed by ion implantation. Hence, present metal/glass composites may be considered similar to be a homogeneous material for light propagation [4]. This is a simplification, which is true generally for low intensity light, but it gives an estimate of the optical penetration depth as (α^{-1}) of the laser pulses into the composite material, where α is the linear absorption coefficient. An intense laser pulse is absorbed and relaxed into heat into the surface SLSG layer of a thickness of α^{-1}, which is several microns [192], i.e., deeper than the thickness of the implanted layer (Fig. 7).

The optical spectral result of pulse laser treatment by 5 pulses of a KrF excimer laser with pulse length of 25 ns full-width at half-maximum at a wavelength of 248 nm with the total released energy of 0.2 J/cm^2 on the optical spectra of the Ag-implanted glass is presented in Fig. 30 [148]. Applied laser pulses did not change the reflectance and transmittance spectra of the non-implanted SLSG, but for implanted sample the location of the transmittance minimum shifts slightly towards shorter wavelengths, and the transmittance in peak position increases from 16 to 23% (Fig. 30a). Remarkable change was found in the reflectance spectra, where in the case of the implanted surface, the peak of the overlapping bands shifts continuously from 490 to 450 nm with modifications in the shape of the envelope of the bands, which become narrower (Fig. 30b). The reflectance intensity falls from 38 to 27 %. However, for reflectance from the rear face of the same sample (Fig. 30c) there is only a decrease of the intensity to a 13 % maximum, which is at the same initial wavelength as in the implanted sample. As presented in Fig. 31, (note an increased magnification of ten times in the direction perpendicular to plane of the figure), it is seen that there are a lot of clearly defined hills (believed to be silver particles) on the glass surface whose size is one order smaller in contrast to large hills on the implanted glass surface in Fig. 29[132].The MNPs accumulate effect on the laser irradiated surface of glass is result of melting implanted composite layer and some desorption of glass material under the laser pulses, which exposes the melted metal particles after their solidification. Hence, the first of the conclusions from the present data on excimer laser treatment of implanted glass is the reduction of the size of the silver nanoparticles, and second is the existence of some asymmetry in their depth distribution (Fig 30b, c). To recognize the mechanisms by which the changes occur for strongly absorbed excimer laser pulses the thermal propagation after the laser-irradiation must be considered. The laser heating is traditionally characterized by the heat diffusion length, $l(t) = (D{\cdot}t)^{1/2}$, where D is the heat diffusivity, and t is the laser pulse duration. In the present experiment with laser pulses of 25 ns, the heat propagation is approximately $l(t) = 115$ nm, that is shorter than the α^{-1}, i.e., the temperature rise is no longer controlled by the diffusion of the heat. However, the $l(t)$ surpasses the depth of the implanted Ag nanoparticles. As was estimated earlier [192], the temperature at the surface of laser treated SLSG reaches values exceeding ~700°C, which is equivalent to the SLSG melting temperature. Under these temperature conditions there is also a possibility for melting small silver particles, because their melting temperature is drastically decreased, for example, to ~400°C for sizes < 30 nm, compared with the bulk melting temperature of 960°C [193]. The time scale of electronic relaxation and energy transfer to the lattice vibrations in the metal particles is several orders faster than in the surrounding glass medium. Therefore, during the interaction of the excimer laser pulse with the metal/glass composite, the Ag nanoparticles are heated and melted more quickly than solidification of the melted glass can occur. Atomically dispersed Ag released from nanoparticles enters into the glass melt, and immediately diffuse throughout all the heated thickness of the laser treated substrate. In principle in time this could lead to a uniform metal distribution, where the silver atom concentration exceeds the solubility value in the solid glass. However, following glass solidification spreading from the depth to the surface, as heat from the laser pulse penetrates into the depth of the sample, the cooling part of the annealing cycle will stimulate new nucleation and re-growth of metal particles. In this case, the possibility of re-growth of metal particles will depend on competition between re-growth and the cooling speed of the moving solidification front, resulting in a new non-uniform size distribution of new MNPs over a depth scale similar to that after ion implantation. Obviously, under some conditions the metal particles may be dispersed into separate metal ions and/or into such small units that they cannot display nanoparticle type optical properties. As shown here, subsequent high power excimer laser pulse treatments of the ion implanted layer may be used to melt, and/or re-grow, the MNPs within the insulator medium. Overall this results in a tighter distribution of small particles. The laser treatments have slightly reduced, but not completely removed evidence for a non-symmetric depth distribution of the silver particles. The Ag-insulator composite material is complex, and so a much wider range of laser pulse conditions, and more data on the cooling rates are required to fully model the changes in the size distributions, which can occur.

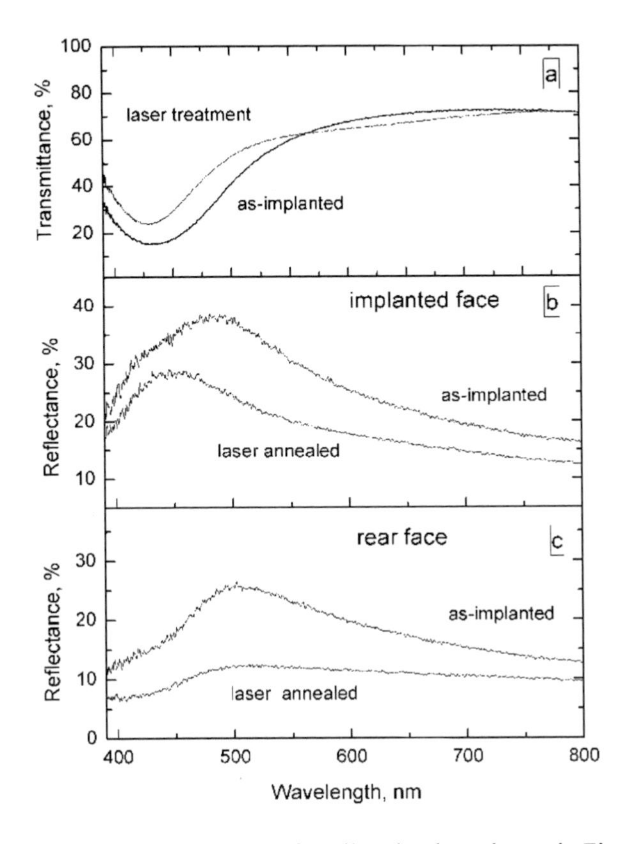

Figure 30. Optical spectra of the SLSG after silver implantation as in Fig. 8 and the implantation followed with laser treatment (0.2 J/cm^2): (a) transmittance; (b) reflectance measured from the implanted face; (c) reflectance measured from rear faces of the sample [148].

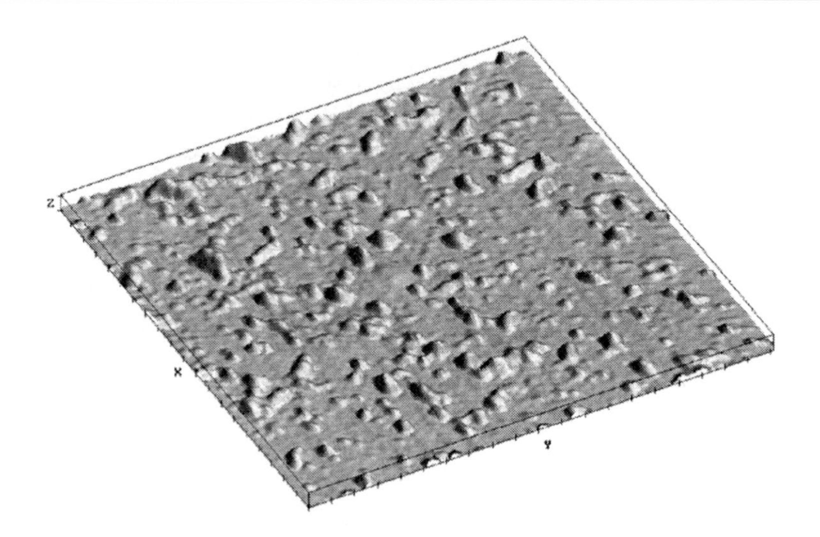

Figure 31. An AFM image as a top view under lateral illumination of the surface of SLSG Ag-implanted with a dose of $7 \cdot 10^{16}$ ion/cm^2 and an energy of 60 keV followed with irradiation with an excimer laser (0.2 J/cm^2). The step along the X and Y axes id 100 nm, and the step along the Z axis is 40 nm [132].

Chapter 11

Ion Synthesis of Silver Nanoparticles in Polymer Matrix

The task of designing new polymer-based composite materials containing MNPs is of current interest. Nanoparticles may be embedded in a polymer matrix in a variety of ways. These are such techniques as chemical synthesis in an organic solvent [1], vacuum deposition on viscous polymers [194], plasma polymerization combined with metal evaporation [195], etc. However, they all suffer from disadvantages, such as a low filling factor or a large distribution in size and shape of nanoparticles, which offsets the good optical properties of composites. Ion implantation is a promising method. Despite the intensive study of MNP synthesis by ion implantation in dielectrics, such as non-organic glasses and crystals, the formation of nanoparticles in organic matrices was realized only at the beginning of the eighties by Koon *et al.* in their experiments on implantation of Fe ions into polymers in 1984 [196]. A first publication on ion-synthesis of noble metal nanoparticles in polymer was realised in 1995 when silver particles were created in PMMA [197]. In Table 2 a comprehensive list of publications on ion synthesis of silver nanoparticles [197-215] with preparation conditions is presented.

The aim of this paragraph is to observe the SPR-related optical absorption of MNPS with an example as silver nanoparticles introduced into polymethylmethacrylate (PMMA) by ion implantation. Substrates 1.2-mm-thick, which are optically-transparent in a wide spectral range (400–1000 nm), were used. PMMA plates were implanted by 30 keV Ag^+ ions with doses in the range from $3.1 \cdot 10^{15}$ to $7.5 \cdot 10^{16}$ ion/cm^2 at ion current density of 4 μA/cm^2. Optical spectra of spherical MNPs embedded in various dielectric media were simulated in terms of the Mie electromagnetic theory [175], which allows one to estimate the extinction cross section σ_{ext} for a light wave incident on a particle.

Table 1. Types of organic matrix with silver nanoparticles synthesized by ion implantation. (Abbreviations – Polycarbonate (PC), Poly(ethyleneterephthalate (PET), Polyimide (PI), Polymethylmethacrylate (PMMA); Phenylmethyl-silane resin with tin diethyldicaprilate (Silicone resin); optical absorption (OA), transmission electron microscopy (TEM), conductivity measurements (CM);room temperature (RT).

Matrix type	Ion energy, keV	Ion dose, ion/cm^2	Current density,μA/cм^2	Matrix temperature,° C	Methods of particle detection	Authors
Epoxy	30	$(0.1\text{-}1.8)\cdot10^{17}$	4	RT	OA, TEM	Stepanov *et al.* 1995 [198] 1997 [199] 2009 [200] Khaibullin I. *et al.* 1997 [201]
PC	60	$3.0\cdot10^{17}$	3	-	OA, TEM	Boldyryeva *et al.* 2004 [202]
PET	40	$(0.1\text{-}2.0)\cdot10^{17}$	4.5	-	CM TEM	Wu *et al.* 2000 [203] 2001 [204, 205] 2002 [206]
PI	130	$(0.1\text{-}5.0)\cdot10^{17}$	1-3	< 350	TEM	Kobayashi *et al.* 2001 [207]
PMMA	30	$(0.1\text{-}7.5)\cdot10^{16}$	4	RT	OA, TEM	Stepanov *et al.* 1994 [197] 2000 [208] 2002 [209] 2004 [210] Bazarov *et al.* 1995 [211]
PMMA	60	$(0.1\text{-}3.0)\cdot10^{17}$	3	-	OA, TEM	Boldyryeva *et al.* 2005 [212]
Silicone resin	30	$(0.1\text{-}1.8)\cdot10^{17}$	4	RT	OA TEM	Khaibullin R. *et al.* 1998 [213] 1999 [214] Stepanov *et al.* 1995 [215]

This value is related to the intensity loss ΔI_{ext} of an incident light beamI_0 passes through a transparent particle-containing dielectric medium due to absorptionσ_{abs} and elastic scatteringσ_{sca}, where $\sigma_{ext} = \sigma_{abs} + \sigma_{sca}$. Following the Lambert-Beer law

$$\Delta I_{ext} = I_0\,(1 - e^{-\#\sigma ext\,h}), \tag{4}$$

where h is the thickness of the optical layer and # - the density of nanoparticles in a sample. The extinction cross section is connected to the extinction constant γ as $\gamma = \#\sigma_{ext}$. Experimental spectral dependencies of optical density (OD) are given by

$$OD = -\lg(I/I_0) = \gamma \cdot \lg(e)h \, , \tag{5}$$

hence, for samples with electromagnetically non-interacting nanoparticles, it possible to put $OD \sim \sigma_{ext}$. Therefore, experimental OD spectra are compared with modeled spectral dependences that are expressed through σ_{ext} calculated by the Mie theory.

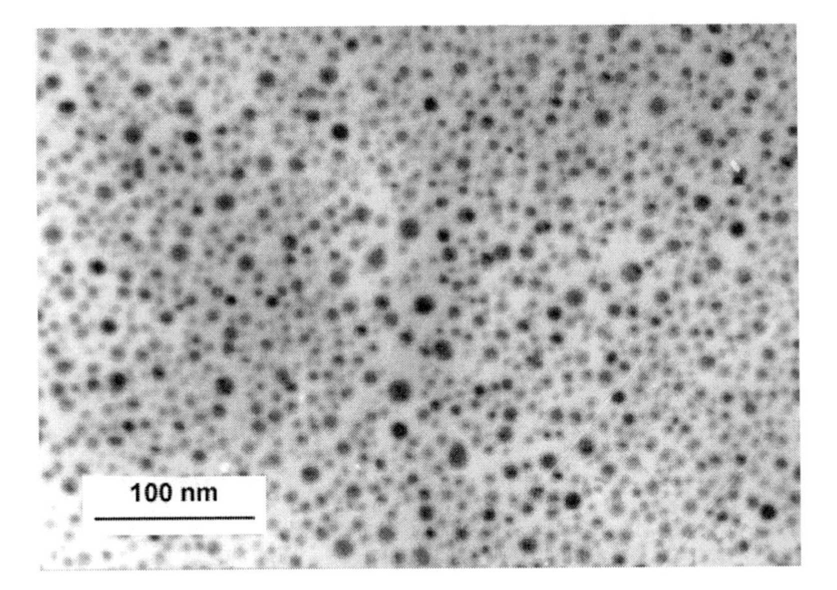

Figure 32. TEM image of silver nanoparticles fabricated in PMMA by Ag-ion implantation [208].

As follows from TEM, Ag-ion implantation results in the formation of silver nanoparticles. As example, the micrograph in Fig. 32 shows spherical nanoparticles synthesized in PMMA at a dose of $5.0 \cdot 10^{16}$ ion/cm^2 [208].Microdiffraction patterns of the composite samples demonstrate that the MNPs have the fcc structure of metallic silver. The diffraction image consists of very thin rings (corresponding to polycrystalline nanoparticles) imposed on wide diffuse faint rings from the amorphous polymer matrix. By comparing the experimental diffraction patterns with standard ASTM data, it possible to conclude that implantation does not form any chemical compounds involving silver ions.

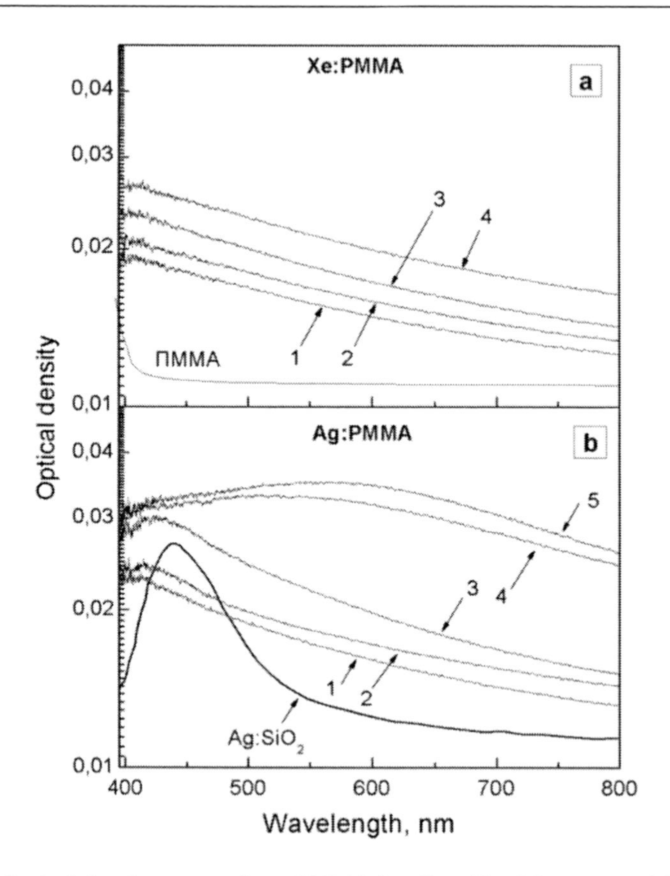

Figure 33. Optical density spectra from PMMA irradiated by (a) xenon and (b) silver ions for doses of (1) $0.3{\cdot}10^{16}$, (2) $0.6{\cdot}10^{16}$, (3) $2.5{\cdot}10^{16}$, (4) $5.0{\cdot}10^{16}$, and (5) $7.5{\cdot}10^{16}$ ion/cm^2. The spectrum taken of SiO_2 implanted by silver ions ($5.0{\cdot}10^{16}$ ion/cm^2) [209].

Optical absorption spectra of PMMA irradiated by xenon and silver ions at various doses are shown in Fig. 33 [208]. As seen in Fig. 33a, when the xenon ion dose increases, the absorption of the polymer in the visible (especially in the close-to-UV) range also increases monotonically. This indicates the presence of radiation-induced structure defects in the PMMA. The implantation by silver ions not only generates radiation defects but also causes the nucleation and growth of MNPs. Therefore, along with the absorption intensity variation as in Fig. 33a, a selected absorption band associated with silver nanoparticles is observed (Fig. 33b). For the lowest ion dose, the maximum of this band is near 420 nm and shifts to red spectral area (up to ~600 nm) with dose increasing, simultaneously with the band broadening. The maximum of this band is not sharp, although it is definitely related to the SPR effect in the silver nanoparticles. Such broad SPR absorption is untypical for silver nanoparticles in PMMA. When silver particles were synthesized in PMMA by the convection melting technique [216], the SPR band was very sharp, unlike present experiment. Fig. 33 shows the *OD* spectrum for inorganic silica glass irradiated by silver ions under the implantation conditions as here. Particle size distributions in the SiO_2 and PMMA are nearly the same. SiO_2 has the refractive index close to that of PMMA. However, the absorption of Ag nanoparticles in the glass (Fig. 33b) is much more narrow and intense than the absorption of the MNPs in the polymer.

The attenuation (extinction) of an optical wave propagating in a medium with MNPs depends on the SPR absorption and the light scattering efficiency. The wavelength of optical radiation, the particle size, and the properties of the environment are governing factors in this process. Within the framework of classical electrodynamics (the Maxwell equations), the problem of interaction between a plane electromagnetic wave and a single spherical particle was exactly solved in terms of optical constants of the selected materials by Mie [175]. The complex value of the optical constant ε_{Ag} [217] and ε_{PMMA} [218] in the visible range was used. The extinction was calculated for particles of size between 1 and 10 nm to be in consistence with experimental sizes (Fig. 32).

As a first step of simulation, consider the simplest case where Ag nanoparticles are incorporated into the PMMA. Simulated extinction spectra of Ag nanoparticles embedded in a polymer matrix shown in Fig. 34 [210]. The extinction feature is a very wide band, which covers the entire spectral range. In the given range of particle sizes, the position of the SPR absorption maximum (near 440 nm) is almost independent of the particle size. In same time, the extinction band intensity grows while the band itself somewhat narrows with increasing particle size. Comparing the modeled and experimental spectra, it is seen that Fig. 34 refers to the situation where PMMA is implanted by silver ions with doses between $0.33 \cdot 10^{16}$ and $2.5 \cdot 10^{16}$ ion/cm^2 (Fig. 34b, curves *1–3*). This dose range corresponds to the early stage of MNP nucleation and growth in the *OD* spectral band with a maximum between 420 and 440 nm. Thus, it is possible to conclude that ion implantation in this dose range results in the formation of Ag nanoparticles, as also revealed by TEM. However, at higher implantation doses, the measured *OD*spectra and the modeled spectra shown in Fig. 34 diverge.

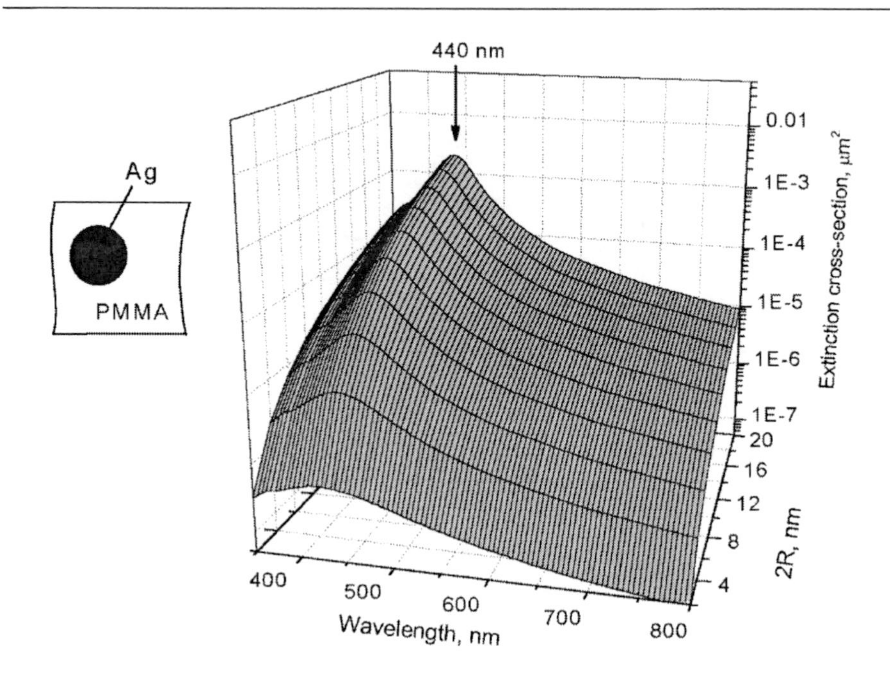

Figure. 34. Simulated optical extinction spectra for silver nanoparticles embedded in PMMA vs. particle size [210].

To explain the experimental dependences corresponding to high-dose silver implantation into PMMA, it should be considered a difference between implantation into polymers and inorganic materials (silicate glasses, crystals, etc.). The most important distinction is that, as the dose increases, so does the number of dangling chemical bonds of polymer along the track of an accelerated ion. Because of this, gaseous hydrogen, low-molecular hydrocarbons (e.g., acetylene), CO, and CO_2 evolve from the matrix [219]. In particular, ion-irradiated PMMA loses HCOOCH3 methoxy groups [220]. The evolution of several organic fractions leads to the accumulation of carbon in the irradiated polymer layer, and radiation-induced chemical processes may cause chain linking. Eventually, an amorphous hydrogenated carbon layer is produced.

To take into account the specific phase structure of the irradiated polymer, it is of interest to analyze the optical properties (extinction) of Ag nanoparticles embedded in the amorphous carbon matrix (C-matrix). For this system, the extinction cross section spectra vs. particle size dependence was simulated (Fig. 35) in the same way as for the MNP–PMMA using complex optical constants $\varepsilon_C$for amorphous carbon [221]. As before (Fig. 34), throughout the particle size interval, the extinction spectra exhibit a single broad band, which covers the visible range, but with a maximum at longer wavelength (~510 nm).

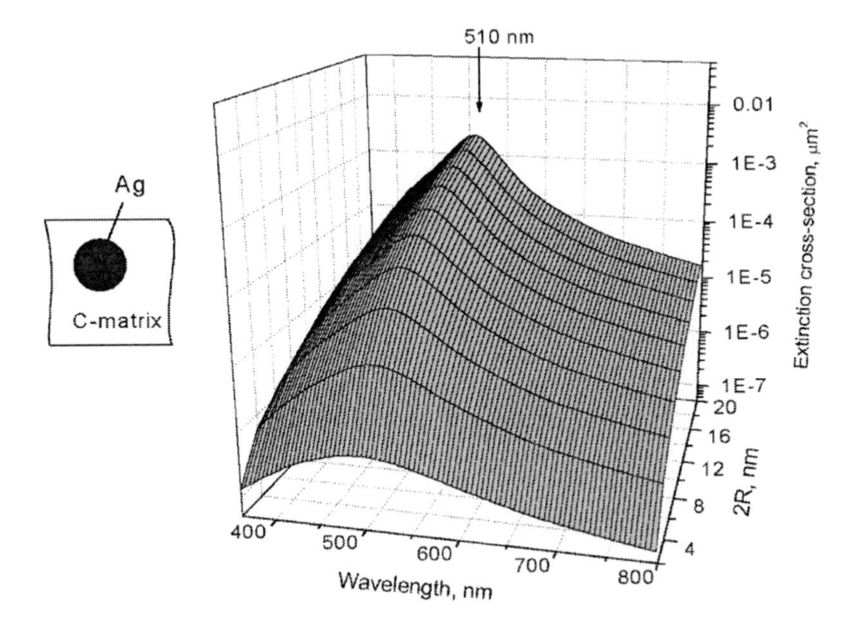

Figure 35. Simulated optical extinction spectra for silver nanoparticles embedded in the C-matrix vs. particle size [210].

This wavelength position of the maximum, which is observed upon changing the matrix, may be assigned to a longer wavelength *OD* band in the experimental spectra for the PMMA, which arises when the Ag ion dose exceeds $2.5 \cdot 10^{16}$ ion/cm^2 (Fig. 34b; curves *3*, *4*). It seems that this spectral shift may be associated with the fact that the pure polymeric environment of the Ag nanoparticles turns into the amorphous carbon as the implantation dose rises. The broader extinction bands in the C-matrix (Fig. 35) compared with the PMMA (Fig. 34) also count in favor of this supposition, since the broadening of the extinction bands is observed in the experiments as well (Fig. 33b). In a number of experiments, however, the carbonization of the polymer surface layer depended on the type of the organic material and accelerated ions, as well as on the implantation parameters, and completed at doses of $(0.5\text{–}5.0) \cdot 10^{16}$ ion/cm^2 but the entire material was not carbonized. The carbon fragments may reach several tens of nanometers in size [219]. Thus, the assumption that the polymer irradiated is completely carbonized, which was used in the simulation (Fig. 35), does not completely correspond to the real situation. Therefore, extinction spectra for nanoparticles represented as a silver core covered by a carbon shell in an insulating matrix (PMMA) will be analyzed in terms of the Mie relationships for shelled cores [222].

Optical extinction spectra for a Ag nanoparticle with a fixed size of the core (4 nm) and a varying thickness of the carbon shell (from 0 to 5 nm) are shown in Fig. 36. The maximum of the SPR bands of the particles is seen to shift from 410 nm (uncovered particle, Fig. 36) to approximately 510 nm. Simultaneously, the SPR band intensity decreases, while the UV absorption increases, so that the absorption intensity at 300 nm and a shell thickness of 5 nm exceeds the SPR absorption of the particles. Both effects (namely, the shift of the SPR band to longer wavelengths and the increased absorption in the near ultraviolet) agree qualitatively with the variation of the experimental *OD* spectra (Fig. 35b) when the implantation dose exceeds $2.5 \cdot 10^{16}$ ion/cm^2. Thus, our assumption that the increase in the carbonized phase fraction with implantation dose and the variation of the *OD* spectra (Fig. 33b) go in parallel is sustained by the simulation of the extinction for complex particles (Fig. 36).

In spite of the fact that the model dependences on the carbon sheath thickness and the experimental dose dependences agree qualitatively, discrepancies still exist, particularly, in the position of the long-wave maximum in the optical density spectra and in the breadths of the simulated and experimental spectra. Possible reasons for such quantitative discrepancies are discussed below.

Interest in carbon-based composites with MNPs goes back a long way. Examples are the studies of magnetic properties of cobalt particles [223], electric and optical properties of layers with copper [224] or silver [225, 226] nanoparticles, etc. It was found in optical absorption experiments that copper and silver nanoparticles [224, 226] dispersed in carbon matrices exhibit a weak SPR effect as in our work (Figs. 33b, 35, 36).

When analyzing the optical properties of nanoparticles embedded in a medium, it should be taking into account effects arising at the particle–matrix interface, such as the static and dynamic redistributions of charges between electronic states in the particles and the environment in view of their chemical constitution [227]. Consider first the charge static redistribution. If an atom is deposited (adsorbed) on the MNP surface(Fig. 37), the energy levels of this

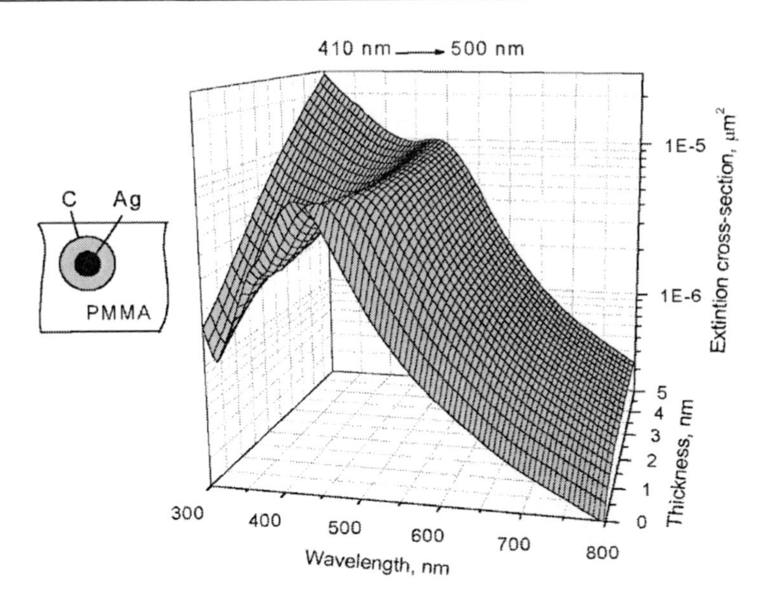

Figure 36. Simulated optical extinction spectra for 4-nm silver nanoparticles with the carbon shell that are placed in the PMMA matrix vs. sheath thickness [210].

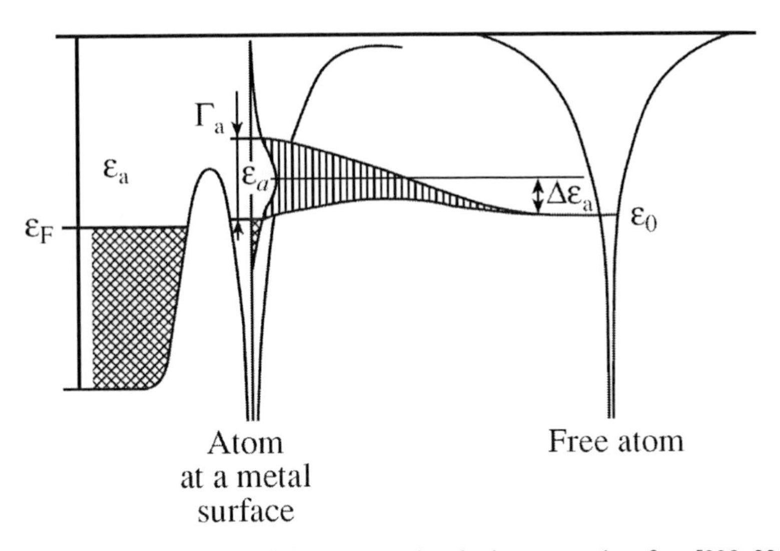

Figure 37. Electron energy levels in an atom absorbed on a metal surface [228, 229].

A free atom (to the right) reches the surface (to the left).). Γ_a is the spread of energy levels ε_a. Electron levels in the conduction band of the metal are occupied up to the Fermi level ε_F.atom ε_a change their positions compared with those in the free state [228,229]. When the number of the adsorbed matrix atoms becomes significant, their contact generates a wide distribution of density of states. Additionally, the adsorbed atoms are separated from surface atoms of the metal by a tunnel barrier. The gap between the energy positions ε_a of the adsorbed atoms and the Fermi level ε_F of the particles depends on the type of the adsorbate. The overlap between the energy positions of the matrix atoms and the energy positions of the silver surface atoms depends on the rate with which the electrons tunnel through the barrier. Accordingly, the conduction electron density in the particles embedded will change compared with that in the particles placed in a vacuum (without adsorbates): it decreases if the electrons tunnel toward the adsorbed atoms or increases when the electrons tunnel in the reverse direction. Eventually, equilibrium between the particle and the matrix sets in; i.e., a constant electrical charge (Coulomb barrier) forms at the nanoparticle surface.

Such a charge static redistribution due to the deposition of an adsorbate on the particle surface and the respective change in the electron concentration in the MNPs could also observed in the SPR absorption spectra [229]. In metals (silver, sodium, aluminum, etc.), where free conduction electrons dominate, the SPR spectral maximum $h\omega_{max}$ depends on the concentrations of electrons, N, and nanoparticles as

$$h\omega_{max} \approx [N/(\varepsilon_0 m_{eff})]^{1/2} [2\varepsilon_m + 1 + \chi_1^{inter}]^{1/2}, \quad (6)$$

where ε_m is the permittivity of the matrix, specifies the contribution of the real part of the susceptibility of interband optical transitions in a metal, and m_{eff} is the effective mass of an electron.

The incorporation of Ag nanoparticles into the carbon matrix of C_{60} fullerene (or the deposition of carbon on the nanoparticle surface) reduces the concentration of 5*sp* electrons in the particle roughly by 20 %, since they are trapped by matrix molecules [230]. According to (10), the decrease in N is bound to shifts the MNP extinction spectrum toward longer wavelength, as was demonstrated by comparing the experimental spectra of the particles in free space (without an adsorbate) with those of the particles in the C_{60} matrix [230].Samples studied in [230] were similar to those obtained by ion implantation in present work (a carbonized layer near silver particles implanted into the polymer). Thus, the shift of the SPR extinction band to the longer wavelength with increasing of implantation dose in present experiment(Fig. 33) may also be explained by the formation of a carbon shell around silver nanoparticles, which traps conduction electrons.

The charge dynamic variation in time at the particle–matrix interface causes the electron concentration in the particle to fluctuate. Fluctuation influences directly to the SPR relaxation. The lifetime of excited conduction electrons in the particle defines the SPR spectral width. Here, the contribution from electron scattering by the interface (because of restrictions imposed on the electron free path [4]) adds up with the charge dynamic variation at the interface. Thus, the temporal capture of conduction electrons from the particle broadens the SPR-related extinction spectra. Such effect was demonstrated with silver nanoparticles embedded in the C_{60} matrix [230]. Silver nanoparticles in the carbon matrix exhibit the much broader SPR band than in free space. We may therefore suppose that, as the dose rises, the charge dynamic redistribution may broaden the SPR spectra of silver nanoparticles synthesized by ion implantation in PMMA. This is because implantation carbonizes the irradiated layer with increasing absorbed dose and raises the amount of acceptor levels on the MNP surface, which changes the relaxation time of electrons excited. The simulation (Fig. 36) also demonstrates the shift of the SPR maximum. However, the effect of charge dynamic redistribution is disregarded in the Mie theory. Therefore, the long-wave shift of the SPR band due to the charge static redistribution at the particle–matrix interface is an additional reason why the experimental spectra are observed at longer waves than the model ones (Figs. 35, 36).

Along with the charge static redistribution at the interface, the charge at the same interface may also change dynamically, i.e., with a high rate. After the static state of the charge has been established and the Fermi level at the interface has been stabilized, the MNP electrons optically excited above the Fermi level (hot electrons) may tunnel (by fluctuations) to the matrix over or through the static barrier (Fig. 37). Levels occupied by the electrons in the intermediate (between the particle and the matrix) state depend on the chemical constitution of the materials. Within a residence lifetime, the electrons may tunnel again from the acceptor levels of the matrix to the particle and this process may occur over and over.

The charge dynamic variation in time at the particle–matrix interface causes the electron concentration in the particle to fluctuate. Fluctuation influences directly the SPR relaxation. The lifetime of excited conduction electrons in the particle defines the SPR spectral width. Here, the contribution from electron scattering by the interface (because of restrictions imposed on the electron free path) adds up with the charge dynamic variation at the interface. Thus, the temporal capture of conduction electrons from the particle broadens the SPR-related extinction spectra. This was demonstrated with a set of silver nanoparticles embedded in the C60 matrix [230]. Silver nanoparticles in the carbon matrix exhibit the much broader SPR band than in free space. We may therefore suppose that, as the dose rises, the charge dynamic redistribution may broaden the SPR spectra of silver nanoparticles synthesized by ion implantation in PMMA. This is because implantation carbonizes the irradiated layer with increasing absorbed dose and raises the amount of acceptor levels on the MNP surface, which change the relaxation time of electrons excited. Since the classical Mie theory disregards the charge dynamic redistribution, the model spectra (Fig. 36) must be narrower than the experimental spectra, which is the case.

In summary, Based on the Mie classical electrodynamic theory, optical extinction spectra for silver nanoparticles in the polymeric or carbon environment, as well as for sheathed particles (silver core +carbon sheath) placed in PMMA, as a function of the implantation dose are simulated. The analytical and experimental spectra are in qualitative agreement. At low doses, simple monatomic silver particles are produced; at higher doses, sheathed particles appear. The quantitative discrepancy between the experimental spectra and analytical spectra obtained in terms of the Mie theory is explained by the fact that the Mie theory disregards the charge static and dynamic redistributions at the particle–matrix interface. The influence of the charge redistribution on the experimental optical spectra taken from the silver–polymer composite at high doses, which cause the carbonization of the polymer irradiated, is discussed.

ACKNOWLEDGMENTS

I wish to thank my partners and co-authors D. Hole, P.D. Townsend, V.I. Nugdin, V.F. Valeev, I.B. Khaibullin, R.I. Khaibullin, V.N. Bazarov, Yu.N. Osin, S.N. Abdullin, V.A. Zikharev, I.A. Faizrakhmanov, A.A. Bukharaev, V.N. Popok, U. Kreibig, A.I. Ryasnyansky, R.A. Ganeev, E. Alves. Also, I grateful to the Alexander von Humboldt Fondation in Germany (I. Physikalisches Institute 1A, der RWTH in Aachen and Laser Zenrum Hannover), Austrian Scientific Foundation in the frame of Lisa Meitner Fellowship (Institute of Physics and Erwin Schrödinger Institute for Nanoscale Research of Karl-Franzens-University in Graz) and the Royal Society/NATO (Sussex University, UK) for financial support. Partly, this work is supported by the Ministry of Education and Science of the Russian Federation (FTP "Scientific and scientific-pedagogical personnel of the innovative Russia" contract No. 02.740.11.0797.

References

[1] Zhang, J. Z. *Optical properties and spectroscopy of nanomaterials*, Wold Sci. Pub.: London, 2009.

[2] Sarychev, A.; Shalaev, V. *Electrodynamics of metamaterials*, Wold Sci. Pub.: New York, 2007.

[3] Haglund, Jr., R. F. In Handbook of optical properties, Vol. II, Optics, of small particles, interfaces, and surfaces; Hummel, R. E.; Wissmann, P.; Eds., CRS Press: New York, 1997.

[4] Kreibig U.; Vollmer M. *Optical Properties of Metal Clusters*, Springer: Berlin, 1995.

[5] Stepanov, A. L.; Khaibullin, I. B.; Townsend, P. D.; Hole, D. E.; Bukharaev, A. A. *Rus. Feder. Patent*, No. 2156490, 2000.

[6] Townsend, P. D.; Massarelli, L. *US Patent*, No. 5102736, 1992.

[7] Hampikian, J. M.; Hunt, E. M. *US Patent*, No. 6294223, 2001.

[8] Kishimoto, N.; Takeda, Y.; Okubo, S.; *Jap. Patent*, No. JP2004091817, 2004.

[9] Townsend, P. T.; Chandler, P. J.; Zhang, L. *Optical Effects of Ion Implantation,* Cambridge Univ. Press: Cambridge, 1994.

[10] Polman, A. *J. Appl. Phys*. 1997 82, 1-39.

[11] Davenas, J.; Perez, A.; Thevenard, P.; Dupuy, C. H. S. *Phys. Stat. Sol. A* 1973, 19, 679-686.

[12] Treilleux, M.; Thevenard, P.; Ghassagne, G.; Hobbs, L. H. *Phys. Stat. Sol. A* 1978, 48, 425-430.

[13] Arnold, G. W. *J. Appl. Phys*. 1975, 46, 4466-4473.

[14] Arnold, G. W.; Borders, J. A. J. Appl. Phys. 1977, 48, 1488-1496.

[15] Stepanov, A. L. In *Metal-Polymer Nanocomposites*; Nicolais, L.; Carotenuto, G.; Eds., John Wiley & Sons Publ: London, 2004, pp. 241-263.

[16] Stepanov, A. L. In *High-power and femtosecond lasers*; Barret, P.-H.; Palmer, M.; Eds., NOVA Sci. Publ. Inc.: New York, 2009, pp. 27-70.

[17] Rahmani, M.; Abu-Hassan, L. H.; Townsend, P. D.; Wilson, I. H.; DestefanisG. L. *Nucl. Inst. Meth. Phys. Res. B* 1988, 32, 56-60.

[18] Rahmani, M.; TownsendP. D. *Vacuum* 1989, 39, 1157-1162.

[19] White, C. W.; Thomas, D. K.; Hensley, D. K.; McCallum, J. C.; Pogany, A.; Haglund Jr., R. F.; Magruder, R. H.; YangL. *Nanostruc. Mat.* 1993, 3, 447-457.

[20] Ila, D.; Williams, E. K.; Sarkisov, S.; Poker, D. B.; HensleyD. K. Mat. *Res. Soc. Symp. Proc.* 1998, 504, 381- 385.

[21] Steiner, G.; Pham, · M. T.; Kuhne, Ch.; SalzerR. Fresenius *J Anal Chem.* 1998, 362, 9-14.

[22] Ganeev, R. A.; Ryasnyanskii, A. I.; Stepanov, A. L.; Marques, C.; da Silva, R. C.; Alves, E. *Opt. Comm.* 2005, 253, 205-213.

[23] Ganeev, R. A.; Ryasnyanskii, A. I.; Stepanov, A. L.; Usmanov, T.; Marques, C.; da Silva, R. C.; Alves E. *Optics and Spectroscopy* 2006, 101, 615-622.

[24] Marques, C.; da Silva, R. C.; Wemans, A.; Maneira, M. J. P.; Kozanecki, A.; Alves E. *Nucl. Inst. Meth. Phys. Res. B* 2006, 242, 104-108.

[25] Mazzoldi, P.; Tramontin, I.; Boscolo-Boscoletto, A.; Battaglin, G.; Arnold G. W. Nucl. Inst. Meth. Phys. Res. B 1993, 80-81, 1192-1196.

[26] Abouchacra, G.; Serughetti J. *Nucl. Inst. Meth. Phys. Res. B* 1986, 14, 282-289.

[27] Fuchs, G.; Abouchacra, G.;Treilleux, M.; Thevenard, P.; Serughetti J. *Nucl. Inst. Meth. Phys. Res. B* 1988, 32, 100-103.

[28] Qian, Y.; Ila, D.; Zimmerman, R. L.; Poker, D. B.; Boatner, L. A.; HensleyD. K. *Nucl. Inst. Meth. Phys. Res. B* 1997, 127, 524-527.

[29] Zimmerman, R. L.; Muntele, C. I.; IlaD. *Surf. Coat. Technol.* 2005, 196, 85-88.

[30] van Huis, M. A.; Fedorov, A. V.; van Veen, A.; Falub, C. V.; Eijt, S. W. H.; Kooi, B. J.; De Hosson, J. Th. M.; Hibma, T.; ZimmermanR. L. *Nucl. Inst. Meth. Phys. Res. B* 2002, 191, 442-446.

[31] Xiao, X. H.; Xu, J. X.; Ren, F.; Liu, C.; JiangC. Z. *Physica E* 2008, 40, 705-708.

[32] Matsunami, N.; Hosono H. *Nucl. Inst. Meth. Phys. Res. B* 1993, 80, 1233-1236.

[33] Deying, S.; Saito, Y.; SuganomataS. *Jpn. J. Appl. Phys*. 1994, 33, L966-L969.

[34] Shang, D. Y.; Saito, Kittaka, Y. R.; Taniguchi, S.; KitaharaA. *J. Appl. Phys.* 1996, 80, 6651-6654.

[35] Saito, Y.; KitaharaA. *J. Appl. Phys*. 2000, 87, 1276-1279.

[36] Fujita, T.; Ijima, K.; Mitsui, N.; Mochiduki, K.; Ho, A. Yi-J.; Saito Y. Jap. *J. Appl. Phys*. 2008, 47, 7224-7229.

[37] Sarkisov, S. S.; Williams, E. K.; Curley, M. J.; Smith, C. C.; Ila, D.; Venkateswarlu, P.; Poker, D. B.; HensleyD. K. *Mat. Res. Soc. Symp. Proc.* 504, 357- 362 (1998).

[38] Sarkisov, S. S.; Williams, E. K.; Curley, M. J.; Ila, D.; Venkateswarlu, P.; Poker, D. B.; HensleyD. K. *Nucl. Inst. Meth. Phys. Res. B* 1998, 141, 294-298.

[39] Sarkisov S. S., Williams E. K., Curley M. J., Smith C. C., Ila, D.; Poker, D. B.; Hensley, D. K.; Banks, C.; PennB. *Proc. SPIE.* 3283, 942-948 (1998).

[40] Sarkisov, S. S.; Williams, E. K.; Curley, M. J.; Ila, D.; Svetchnikov, V. L.; Pan, V. M.; Poker, D. B.; Hensley, D. K.; Banks, C.; Penn, B.; WangJ. W. *Proc. SPIE.* 1998, 3413, 98-110.

[41] S. S. Sarkisov, M. J. Curley, E. K. Williams, D. Ila, V. L. Svetchnikov, H. W. Zandbergen, G. A. Zykov, D. B. Poker, D. K. Hensley*Proc.SPIE.* 1999, 3790, 43-55.

[42] Sarkisov, S. S.; Curley, M. J.; Williams, E. K.; Ila, D.; Svetchnikov, V. L.; Zandbergen, H. W.; Zykov, G. A.; Banks, C.; Wang, J.-C.; Poker, D. B.; HensleyD. K. *Nucl. Inst. Meth. Phys. Res. B* 2000, 166-167, 750-757.

[43] Williams, E. K.; Ila, D.; Sarkisov, S. S.; Curley, M. J.; Poker, D. B.; Hensley, D. K.; BorelC. *Mat. Res. Soc. Symp. Proc*. 1998, 504, 363-369.

[44] Williams, E. K.; Ila, D.; Sarkisov, S. S.; Curley, M. J.; Cochrane, J. C.; Poker, D. B.; Hensley, D. K.; BorelC. *Nucl. Inst. Meth. Phys. Res. B* 1998, 141, 268-273.

[45] Williams, E. K.; Ila, D.; Darwish, A.; Poker, D. B.; Sarkisov, S. S.; Curley, M. J.; Wang, J.-C.; Svetchnikov, V. L.; ZandbergenH. W. *Nucl. Inst. Meth. Phys. Res. B* 1999, 148, 1074-1078.

[46] Amolo, G. O.; Comins, J. D.; Naidoo, S. R.; Connell, S. H.; Witcomb, M. J.; DerryT. E. *Nucl. Inst. Meth. Phys. Res. B* 1999, 250, 233-237.

[47] Mazzoldi, P.; Tramontin, L.; Boscolo-Boscoletto, A.; Battaglin, G.; Arnold G. W. *Nucl. Inst. Meth. Phys. Res. B* 1993, 80-81, 233-237.

[48] Mazzoldi, P.; Mattei G. *Rivista del nuovo cimento* 2005, 28, 1-69.

[49] Mazzoldi, P.; MatteiG. *Phys. Stat. Sol. A* 2007, 204, 621-630.

[50] Antonello, M.; Arnold, G. W.; Battaglin, G.; Bertoncello, R.; Cattaruzza, E.; Colombo, P.; Mattei, G.; Mazzoldi, P.; TrivillinF. *J. Mater. Chem.* 1998, 8, 457-461.

[51] Battaglin, G.; Cattaruzza, E.; D'Acapito, F.; Gonella, F.; Mazzoldi, P.; Mobilio, S.; PrioloF. *Nucl. Inst. Meth. Phys. Res. B* 1998, 141, 252-255.

[52] Battaglin, G.; Catalano, M.; Cattaruzza, E.; D'Acapito, F.; De Julian Fernandez, C.; De Marchi, G.; Gonella, F.; Mattei, G.; Maurizio, C.; Mazzoldi, P.; Miotello, A.; SadaC. *Nucl. Inst. Meth. Phys. Res. B* 2001, 178, 176-179.

[53] Bertoncello, R.; Gross, S.; Trivillin, F.; Cattaruzza, E.; Mattei, G.; Caccavale, F.; Mazzoldi, P.; BattaglinG. *J. Mater. Res.* 1999, 14, 2449-2457.

[54] CaccavaleF. Pramana - *J. Phys*. 1998, 50, 653-668.

[55] Cattaruzza, E.; Battaglin, G.; Polloni, R.; Cesca, T.; Gonella, F.; Mattei, G.; Maurizio, C.; Mazzoldi, P.; D'Acapito, F.; Zontone, F.; BertoncelloR. *Nucl. Inst. Meth. Phys. Res. B* 1990, 148, 1007-1011.

[56] Gonella, F.; Mattei, G.; Mazzoldi, P.; Sada, C.; Battaglin, G.; Cattaruzza E. *Appl. Phys. Lett*. 1999, 75, 55-57.

[57] Matsunami, N.; HosonoH. *Appl. Phys. Lett*. 1993, 63, 2050-2053.

[58] Magruder III, R. H.; Zuhr, R. A.; Osborne, Jr.D. H. *Nucl. Inst. Meth. Phys. Res. B* 1995, 99, 590-593.

[59] Magruder III, R. H.; Anderson, T. S.; Zuhr, R. A.; ThomasD. K. Nucl. *Inst. Meth. Phys. Res. B* 1996, 108, 305-312.

[60] Magruder III, R. H.; Robinson, S. J.; Smith, C.; Meldrum, A.; Halabica, A.; Haglund, Jr.,R. H. *J. Appl. Phys*. 2009, 105, 24303-8.

[61] Anderson, T. S.; Magruder III, R. H.; Zuhr, R. A.; WittigJ. E. *J. Electronic Mater*. 1996, 25, 27-35.

[62] Anderson, T. S.; Magruder III, R. H.; Kinser, D. L.; Zuhr, R. A.; ThomasD. K. *Nucl. Inst. Meth. Phys. Res*. B 1997, 124, 40-46.

[63] Anderson, T. S.; Magruder III, R. H.; Kinser, D. L.; Wittig, J. E.; Zuhr, R. A.; ThomasD. K. *J. Non.-Cryst. Sol.* 1998, 224, 299-306.

[64] Anderson, T. S.; Magruder III, R. H.; Wittig, J. E.; Kinser, D. L.; ZuhrR. A. *Nucl. Inst. Meth. Phys. Res. B* 2000, 171, 401-405.

[65] Zuhr, R. A.; Magruder III, R. H.; AndersonT. S. *Surf. Coat. Technol.* 1998, 101, 401-408.

[66] Pham, M. T.; Matz, W.; SeifarthH. *Anal. Chim. Acta* 1997, 350, 209-220.

[67] Liu, Z.; Wang, H.; Li, H.; WangX. *Appl. Phys. Lett.* 1998, 72, 1823-1825.

[68] Liu, Z.; Li, H.; Feng, X.; Ren, S.; Liu, Z.; LuB. *J. Appl. Phys*. 1998, 84, 1913-1917.

[69] Liu, Z.; Li, H.; Wang, H.; Shen, D.; Wang, X.; AlkemadeP. F. A.*J. Mater. Res*. 2000, 15, 1245-1247.

[70] Ila, D.; Williams, E. K.; Sarkisov, S.; Smith, C. C.; Poker, D. B.; HensleyD. K. *Nucl. Inst. Meth. Phys. Res*. B 1998, 141, 289-293.

[71] D'Acapito, F.; ZontoneF.J. Appl. Cryst. 1999, 32, 234-240.

[72] Stepanov, A. L.; Hole, D. E.; Townsend P. D. *Nucl. Inst. Meth. Phys.* Res. B 2000, 166-167, 882-886.

[73] Stepanov A. L. Rev. *Adv. Mater. Sci*. 2003, 4, 45-60.

[74] Jiang, C. Z.; FanX. *J. Surf. Coat. Technol*. 2000, 131, 330-333.

[75] Ren, F.; Jiang, C. Z.; Chen, H. B.; Shi, Y.; Liu, C.; WangJ. B. *Physics B* 2004, 353, 92-97.

[76] Ren, F.; Jiang, C. Z.; Zhang, L.; Shi, Y.; Wang, J. B.; WangR. H. *Micron* 2004, 35, 489-493.

[77] Ren, F.; Jiang, C. Z.; Liu, C.; ShiY. *J. Kor. Phys. Soc*. 2005, 46, S43-S47.

[78] Ren, F.; Jiang, C. Z.; Fu, D. J.; RuQ. Jap. J. Appl. Phys. 2005, 44, 8512-8514.

[79] Ren, F.; Jiang, C. Z.; Liu, C.; Fu, D.; ShiY. *Solid State Comm*. 2005, 135, 268-272.

[80] Ren, F.; Jiang, C. Z.; Liu, C.; Wang, J.; OkuT. *Phys. Rev. Lett*. 2006, 97, 165501-1 - 4.

[81] Ren, F.; Jiang, C. Z.; Cai, G. X.; Fu, Q.; ShiY. *Nucl. Inst. Meth. Phys. Res. B* 2007, 262, 201-204.

[82] Ren, F.; Cai, G. X.; Xiao, X. H.; Fan, L. X.; Liu, C.; Fu, D. J.; Wang, J. B.; JiangC. Z. *J. Appl. Phys*. 2008, 103, 843308-1 - 5.

[83] Ren, F.; Xiao, X. H.; Cai, G. X.; Wang, J. B.; JiangC. Z. *Appl. Phys. A* 2009, 96, 317-325.

[84] Liu, X .F.; Jiang, C. Z.; Feng, R.; FuQ. *Acta Phys. Sinca* 2005, 54, 4633-4637.

[85] Xiao, X. H.; Jiang, C. Z.; Ren, F.; Wang, J.; ShiY. *Solid State Comm.* 2006, 137, 362-365.

[86] Xiao, X. H.; Ren, F.; Wang, J. B.; Liu, C.; JiangC. Z. *Mater. Lett*. 2007, 61, 4435-4437.

[87] Xiao, X. H.; Guo, L. P.; Ren, F.; Wang, J. B.; Fu, D. J.; Chen, D. L.; Wu, Z. Y.; Jia, Q. J.; Liu, C.; JangC. Z. *Appl. Phys. A* 2007, 89, 681-684.

[88] Wang, Y. H.; Jiang, C. Z.; Ren, F.; Wang, Q. Q.; Chen, D. J.; FuD. J. *J. Mater. Sci*. 2007, 42, 7294-7298.

[89] Wang, Y. H.; Jiang, C. Z.; Xiao, X. H.; ChenD. *J. Physica B* 2008, 403, 2143-2147.

[90] Zhang, L.; Jiang, C. Z.; Ren, F.; Chen, H.-B.; Shi, Y.; FuQ. *Acta Phys. Sinica* 2004, 53, 2910-2914.

[91] Cai, G. X.; Ren, F.; Xiao, X. H.; Fan, L. X.; JiangC. Z. *Nucl. Inst. Meth. Phys. Res. B* 2008, 266, 889-893.

[92] Cai, G. X.; Ren, F.; Xiao, X. H.; Fan, L. X.; Zhou, X. D.; Jiang C. Z. *J. Mater. Sci. Technol.* 2009, 25, 669-672.

[93] Armelao, L.; Bertoncello, R.; Cattaruzza, E.; Gialanella, S.; Gross, S.; Mattei, G.; Mazzoldi, P.; TondelloE. *J. Mater. Chem.* 2002, 12, 2401-2407.

[94] Ishikawa, J.; Tsuji, H.; Motono, M.; Gotoh, Y.; Arai, N.; Adachi, K.; KotakiH. *IEEE* 2002, 12, 690-693.

[95] Ishikawa, J.; Tsuji, H.; Motono, M.; GotohY. *Surf. Coat. Technol.* 2009, 203, 2351-2356.

[96] Tsuji, H.; Kurita, K.; Gotoh, Y.; Kishimoto, N.; IshikawaJ. *Nucl. Inst. Meth. Phys. Res. B* 2002, 195, 315-319.

[97] Tsuji, H.; Kurita, K.; MotonoM.J. *Vac. Soc. Jap.* 2002, 45, 528-532.

[98] Tsuji, H.; Arai, N.; Motono, M.; Gotoh, Y.; Abachi, K.; Kotaki, H.; IshikawaJ. *Nucl. Inst. Meth. Phys. Res. B* 2003, 206, 615-619.

[99] Tsuji, H.; Arai, N.; Matsumoto, T.; Ueno, K.; Gotoh, Y.; Abachi, K.; Kotaki, H.; IshikawaJ. *Appl. Surf. Sci.* 2004, 238, 132-137.

[100] Tsuji, H.; Arai, N.; Matsumoto, T.; Ueno, K.; Abachi, K.; Kotaki, H.; Gotoh, Y.; IshikawaJ. *Surf. Coat. Technol.* 2005, 196, 39-43.

[101] Arai, N.; Tsuji, H.; Motono, M.; Goto, Y.; Adachi, K.; Kotaki, H.; IshikawaJ. *Nucl. Inst. Meth. Phys. Res. B* 2003, 206, 629-633.

[102] Arai, N.; Tsuji, H.; Ueno, K.; Matsumoto, T.; Gotoh, Y.; Abachi, K.; Kotaki, H.; IshikawaJ. *Surf. Coat. Technol.* 2005, 196, 44-49 (2005).

[103] Arai, N.; Tsuji, H.; Ueno, K.; Matsumoto, T.; Gotoh, Y.; Abachi, K.; Kotaki, H.; Gotoh, Y.; IshikawaJ. *Nucl. Inst. Meth. Phys. Res.* B 2006, 242, 217-220.

[104] Arai, N.; Tsuji, H.; Abachi, K.; Kotaki, H.; Gotoh, Y.; IshikawaJ. Jap. *J. Appl. Phys.* 2007, 46, 6260-6266.

[105] Arai, N.; Tsuji, H.; Gotoh, N.; Matsumoto, T.; Ishibashi, T.; Adachi, K.; Kotaki, H.; Gotoh, Y.; IshikawaJ. *J. Phys.: Conf. Ser.* 2007, 61, 41-45.

[106] Roiz, J.; Oliver, A.; Munoz, E.; Rodríguez-Fernández, L.; Hernandez, J. M.; Cheang-WongJ. C. *J. App. Phys.* 2004, 95, 1783-1791.

[107] Oliver, A.; Reyes-Esqueda, J. A.; Cheang-Wong, J. C.; Román-Velázquez, C. E.; Crespo-Sosa, A.; Rodríguez-Fernández, L.; Seman, J. A.; Noguez C. *Phys. Rev. B* 2006, 74, 245425-1 – 6.

[108] Cheang-Wong, J. C.; Oliver, A.; Rodríguez-Fernández, L.; Arenas-Alatorre, J.; Peña, O.; Crespo-SosaA. *Revista Mexicana de Física S* 2007, 53, 49-54.

[109] Peña, O.; Cheang-Wong, J. C.; Rodríguez-Fernández, L.; Arenas-Alatorre, J.; Crespo-Sosa, A.; Rodríguez -Iglesias, V.; OliverA. *Nucl. Inst. Meth. Phys. Res. B* 2007, 257, 99-103.

[110] Peña, O.; Pal, U.; Rodríguez-Fernández, L.; Silva-Pereyra, H. G.; Rodríguez -Iglesias, V.; Cheang-Wong, J. C.; Arenas-Alatorre, J.; Oliver A. *J. Phys. Chem.* 2009, 113, 2296-2300.

[111] Reyes-Esqueda, J. A.; Torres-Torres, C.; Cheang-Wong, J. C.; Crespo-Sosa, A.; Rodríguez-Fernández, L.; Noguez, C.; Oliver A. *Opt. Express* 2008, 16, 710-713.

[112] Reyes-Esqueda, J. A.; Rodríguez -Iglesias, V.; Silva-Pereyra, H. G.; Torres-Torres, C.; Santiago-Ramirez, A.-L.; Cheang-Wong, J. C.; Crespo-Sosa, A.; Rodríguez-Fernández, L.; Lopez-Suarez, A.; Oliver A. *Opt. Express* 2009, 17, 12849-12868.

[113] Rodríguez -Iglesias, V.; Silva-Pereyra, H. G.; Cheang-Wong, J. C.; Reyes-Esqueda, J. A.; Rodríguez-Fernández, L.; Crespo-Sosa, A.; Kellerman, G.; OliverA. *Nucl. Inst. Meth. Phys. Res. B* 2008, 266, 3138-3142.

[114] Rodríguez -Iglesias, V.; Silva-Pereyra, H. G.; Torres-Torres, C.; Reyes-Esqueda, J. A.; Cheang-Wong, J. C.; Crespo-Sosa, A.; Rodríguez-Fernández, L.; López-Suárez, A.; Oliver A. *Opt. Comm.* 2009, 282, 4157-4161.

[115] Rangel-Rojo, R.; McCarthy, J.; Bookey, H. T.; Kar, A. K.; Rodríguez-Fernández, L.; Cheang-Wong, J. C.; Crespo-Sosa, A.; Lopez-Sosa, A.; Oliver, A.; Rodríguez -Iglesias, V.; Silva-Pereyra H. G. *Opt. Comm.* 2009, 282, 1909-1912.

[116] Romanyuk, A.; Spassov, V.; Melnik V. *J. Appl. Phys.* 2006, 99, 034314-1 - 4.

[117] Takeda, Y.; Plaksin, O.A.; Lu, J.; KishimotoN. *Vacuum* 2006, 80, 776-779.

[118] Joseph, B.; Suchan Sandeep, C. S.; Sekhar, B. R.; Mahapatra, D. P.; PhilipR. *Nucl. Inst. Meth. Phys. Res. B* 2007, 265, 631-636.

[119] Joseph, B.; Lenka, H. P.; Kuiri, P. K.; Mahapatra, D. P.; KesavamoorthyR. Intern. J. *Nanoscience* 2007, 6, 423-430.

[120] Sahu, G.; Rath, S. K.; Joseph, B.; Roy, G. S.; Mahapatra D. P. *Vacuum* 2009, 83, 836-840.

[121] Carles, R.; Farcău, C.; Bonafos, C.; Benassayag, G.; Pécassou, B.; ZwickA. *Nanotechnol.* 2009, 20, 1-6.

[122] Wang, Y. H.; Peng, S. J.; Lu, J. D.; Wang, R. W.; Mao Y. I.; ChenY. G. *Vacuum* 2009, 83, 408-411.

[123] Y. H. Wang, S. J. Peng, J. D. Lu, R. W. Wang, Y. G. Chen, Y.I. Mao*Vacuum* 2009, 83, 412-415.

[124] Magruder III, R.H.; Meldrum A. J. *Non.-Cryst. Solids* 2007, 353, 4813-4818.

[125] Nistor, L. C.; von Landuyt, J.; Barton, J. B.; Hole, D. E.; Skelland, N. D.; Townsend P. D. J. *Non.-Cryst. Solids* 1993, 162, 217-224.

[126] Wood, R. A., Townsend, P. D., Skelland, N. D., Hole, D. E., Barton, J., AfonsoC. N. *J. Appl. Phys*. 1993, 74, 5754-5756.

[127] Dubiel, M., Hofmeister, H., SchurigE. *Phys. Stat. Sol.* B 1997, 203, R5-R6.

[128] Dubiel, M.; Hofmeister, H.; Schurig, E.; Wendler, E.; WeschW. *Nucl. Inst. Meth. Phys. Res. B* 2000, 166-167, 871-876.

[129] Dubiel, M.; Hofmeister, H.; Tan, G. L.; Schicke, K.-D.; WendlerE. *Eur. Phys. J. D* 2003, 24, 361-364.

[130] Dubiel, M.; Hofmeister, H.; Wendler E. J. *Non.-Cryst. Solids* 2008, 354, 607-611.

[131] Seifert, G.; Stalmashonak, S.; Hofmeister, H.; Haug, J.; Dubiel M. *Nanoscale Res. Lett.* 2009, 4, 1380-1383.

[132] Stepanov, A. L.; Hole, D. E.; Bukharaev, A. A.; Townsend, P. D.; N. I. *Nurgazizov Appl. Surf. Sci*. 1998, 136, 298-305.

[133] Stepanov, A. L.; Hole, D. E.; Townsend P. D. J. *Non.-Cryst. Solids* 1999, 224, 275-279

[134] Stepanov, A. L.; Hole, D. E.; Townsend P. D. J. *Non.-Cryst. Solids* 1999, 260, 67-74

[135] Stepanov, A. L.; Hole, D. E.; Townsend P. D. *Nucl. Inst. Meth. Phys. Res. B* 1999, 149, 89-98.

[136] Stepanov, A. L.; Hole, D. E.; Townsend P. D. *Nucl. Inst. Meth. Phys. Res. B* 2000, 161-163, 913-916.

[137] Stepanov, A. L.; Zhikharev, V. A.; Hole, D. E.; Townsend, P. D.; Khaibullin I. B. *Nucl. Inst. Meth. Phys. Res. B* 2000, 166-167, 26-30.

[138] Stepanov, A. L.; Hole, D. E.; Townsend P. D. *Nucl. Inst. Meth. Phys. Res. B* 2000, 166-167, 882-886.

[139] Stepanov A. L. *Optics and Spectroscopy* 2000, 89, 408-412.

[140] Stepanov, A. L.; Popok, V. N.; Hole, D. E.; Bukharaev A. A. *Physics of Solid State* 2001, 43, 2192-2198

[141] Stepanov, A. L.; Popok V. N. J. *Appl. Spectr.* 2001, 68, 164-169

[142] Stepanov, A. L.; Hole, D. E.; Popok V. N. *Tech. Phys. Lett.* 2001, 27, 554-556

[143] Stepanov, A. L.; Hole, D. E.; Bukharaev A. A. *Vacuum*, 2002, 64, 169-177.

[144] Stepanov, A. L.; Hole, D. E.; Popok V. N. *Glass Phys. Chem.* 2002, 28, 90-95.

[145] Stepanov, A. L.; Hole D. E. *Phil. Mag. Lett.* 2002, 82, 149-155.

[146] Stepanov A. L. *In Recent research development in applied physics*; PandalaiS. G.; Ed.; Transworld Res. Network: Kerala, 2002, 5, 1-25.

[147] Stepanov, A. L.; Popok V. N. *Tech. Phys. Lette.* 2003, 29, 977-979.

[148] Stepanov A. L. *Rev. Adv. Mater. Sci* 2003, 4, 45-60.

[149] Stepanov A. L. In *Recent research development in non.-crystalline solids*; PandalaiS. G.; Ed.; Transworld Res. Network: Kerala, 2003, 3, 177-193.

[150] Stepanov, A. L.; Popok V. N. *Surf. Coat. Thechnol.* 2004, 185, 30-37.

[151] Stepanov, A. L.; Popok V. N. *Surf. Sci.* 2004, 566-568, 1250-1254.

[152] Stepanov, A. L.; Popok V. N. *J. Appl. Spectr.* 2005, 72, 229-234.

[153] Stepanov, A. L.; Chichkov, B. N.; Valeev, V. F.; Faizrakhmanov I. A. *Tech. Phys. Lett.* 2008, 34, 184-186.

[154] Stepanov, A. L.; Valeev, V. F.; Faizrakhmanov, I. A.; Chichkov B. N. Tech. *Phys. Lett.* 2008, 34, 1014-1017.

[155] L. Stepanov, V. F. Valeev, V. V. Bazarov, I. A. Faizrakhmanov Excimer laser-assisted annealing of silicate glass with ion-synthesised silver nanoparticles*Tech. Phys.*2009, 54, 1504-1510.

[156] Tsuji, H.; Sugahara, H.; Gotoh, Y.; IshikawaJ..*Surf. Coat. Thechnol.* 2002, 158-159, 208-213.

[157] Tsuji, H.; Sugahara, H.; Gotoh, Y.; IshikawaJ. *Nucl. Inst. Meth. Phys. Res. B* 2003, 206, 249-253.

[158] Tsuji, H.; Sakai, N.; Sugahara, H.; Gotoh, Y.; IshikawaJ. *Nucl. Inst. Meth. Phys. Res. B* 2005, 237, 433-437.

[159] Tsuji, H.; Sakai, N.; Gotoh, Y.; IshikawaJ. *Nucl. Inst. Meth. Phys. Res. B* 2006 242, 129-132.

[160] Saito, Y.; Imamura, Y.; Kitahara A. *Nucl. Inst. Meth. Phys. Res. B* 2003, 205, 272-276.

[161] Fujita, T.; Ijima, K.; Mitsui, N.; Mochiduki, K.; Saito Y. Jap. *J. Appl. Phys.* 2007, 46, 7362-7364.

[162] Ziegler, J. F.; Biersack, J.P.; Littmark, U. *The stopping and range of ions in solids*, Pergamon Press: New York, 1996.
[163] Nastasi, M.; Mayer, J. W.; Hiroven, J.K. *Ion-solid interactions: Fundamentals and applications*, Cambridge Univ. Press: Cambridge, 1996.
[164] Namba, S.; Masuda, K., Gamo, K., Doi, A.; Ishinara, S., Kimura, I. Ion implantation. Proc. Of conf. on ion implantation in semiconductors; Eisen, F. D.; Chadderton, L. T.; Eds.; *Cordon and Breach Sci. Publ.*: London, 1971, pp. 231-236.
[165] Berger, A. J. *Non.-Cryst. Solids* 1992, 151, 88-94.
[166] Skelland, N. D.; Townsend P. D. J. *Non.-Cryst. Solids* 1995, 188, 243-253.
[167] Hole, D.E.; Stepanov, A.L.; Townsend, P.D. *Nucl. Inst. Meth. Phys. Res. B* 1999, 148, 1054-1058.
[168] Konoplev, V. M. Radiat. *Eff. Lett.* 1986, 87, 207-214.
[169] Konoplev, V. M.; Vicanek, M.; Gras-Martí, A. *Nucl. Inst. Meth. Phys. Res. B* 1992, 67, 574-579.
[170] Stepanov, A. L.; Abdullin, S. N.; Khaibullin, R. I.; Valeev, V. F; Osin, Yu. N.; Bazarov, V. V.; Khaibullin, I. B. *Mat. Res. Soc. Proc*. 1995, 392, 267-272.
[171] Khaibullin, R. I.; Osin, Yu. N.; Stepanov, A. L.; Khaibullin, I. B. *Nucl. Inst. Meth. Phys. Res. B* 1999, 148, 1023-1028.
[172] Townsend, P. D.; Can, N.; Chandler, P. J.; Farmery, B. W.; Lopez-Heredero, R.; Peto, A.; Salvin, L.; Underdown, D.; Yang, B. J. *Non.-Cryst. Solids* 1998, 223, 73-85.
[173] Wang, P.W. *Appl. Surf.. Sci.* 1997, 120, 291-298.
[174] Bartels, J.; Lembke, U.; Pascova, R.; Schmelzer, J., Gustzow, I. J. *Non.-Cryst. Solids* 1991, 136, 181-197.
[175] Mie, G. *Ann. Phys*. 1908, 25, 377-445.
[176] Heavens, O. S. *Optical properties of thin solid films, Butterworths Sci.* Pub.: London, 1955.
[177] Maxwell Garnet, J. C. *Philos. Trans. R. Soc. Lond*. 1904, 203, 385-420.
[178] Maxwell Garnet, J. C. *Philos. Trans. R. Soc. Lond.* 1906, 205 (1906) 237-288.
[179] Faik, A.; Allen, L.; Eicher, C.; Gagola, A.; Townsend, P. D.; Pitt, C. W. *J. Appl. Phys*. 1983, 54, 2597-2601.
[180] Hosono, H. Jpn. *J. Appl. Phys.* 1993, 32, 3892-3894.
[181] Palpant, B. *In Non-linear optical properties of matte*; Papadopoulos, M. G.; Ed.; Springer: Berlin, 2006, pp. 461-508.

[182] Hamanaka, Y.; Hayashi, N., Nakamura, A.; Omi, S. J. *Lumines*. 2000, 87-89, 859-861.

[183] Ganeev, R. A.; Ryasnyanskii, A. I.; Stepanov, A. L.; Kondirov, M. K.; Usmanov, T. *Optics and Spectroscopy* 2003, 95, 1034-1042.

[184] Ganeev, R. A.; Ryasnyanskii, A. I.; Stepanov, A. L.; Usmanov, T. *Opt. Quant. Electr.* 2004, 36, 949-960.

[185] Kurata, H.; Takami, A.; Koda, S. *Appl. Phys. Lett.* 1998, 72, 789-791.

[186] Link, S.; Burda, C; Mohamed, M. B.; Nikoobakht, B.; El-Sayed, M. A. *J. Phys. Chem.* A 1999, 103, 1165-1170.

[187] Mafune, F.; Kohno, J.; Takeda, Y.; Kondow, T. J. *Phys. Chem. B* 2002, 106, 8555-8561.

[188] Chandrasekharan, N.; Kamat, P. V., Hu, J.; Jones II, G. J. *Phys. Chem. B* 2000, 104, 11103-11109.

[189] Osborne Jr., D. H.; Haglund Jr., R. F.; Gonella, F.; Garrido, F. *Appl. Phys. B* 1998, 66, 517-521.

[190] Kyoung, M.; Lee, M. *Opt. Comm.* 1999, 171, 145-148.

[191] Buerhop, C.; Blumenthal, B.; Weissmann, R. *Appl. Surf. Sci.* 1999, 46, 653-668.

[192] Townsend, P.N.; Olivares, J. *Appl. Surf. Sci.* 1997, 109-110, 275-282.

[193] Castro, T.; Reifenberger, R.; Choi, T.; Andres, P. P. *Phys. Rev.* B 1990, 42, 8549-8556.

[194] Stepanov, A. L.; Valeev, V. F.; Osin, Yu. N.; Nuzhdin, V. I.; Faizrakhmanov, I. A. *Tech. Phys*. 2009, 54, 997-1001.

[195] Heilmann, A. *Polymer films with embedded metal nanoparticles*, Springer: Berlin, 2003.

[196] Koon, N. C.; Weber, D.; Pehrsson, P.; Sindler, A. I. *Mater. Res. Soc. Proc*. 1984, 27, 445-449.

[197] Stepanov, A. L.; Abdullin, S. N.; Khaibullin, R.I.; Osin, Yu. N.; Khaibullin, I. B. *Proc. Royal Micr. Sci*. 1994, 29, 226.

[198] Stepanov, A. L.; Abdullin, S. N.; Khaibullin, R. I.; Valeev, V. F.; Osin, Yu. N.; Bazarov, V. V.; Khaibullin, I. B. *Mat. Res. Soc. Proc*. 1995, 392, 267-272.

[199] Stepanov, A. L.; Abdullin, S. N.; Khaibullin, R. I.; Khaibullin, I. B. *Patent PF* 109708, 1997.

[200] Stepanov, A. L.; Khaibullin, R. I.; Valeev, V. F.; Osin, Yu. N.; Nuzhdin, V. I.; Faizrakhmanov, I. A. *Tech. Phys*. 2009, 54, 1162-1167.

[201] Khaibullin, I. B.; Khaibullin, R. I.; Abdullin, S. N.; Stepanov, A. L.; Osin, Yu. N.; Bazarov, V. V.; Kurzin, S. P. *Nucl. Inst. Meth. Phys. Res.* B 1997, 127-128, 685-690.

[202] Boldyryeva, H.; Kishimoto, N.; Umeda, N.; Kono, K.; Plaksin, O. A.; TakedaY. *Nucl. Inst. Meth. Phys. Res.* B 2004, 219-220, 953-958.
[203] Wu, Y.; Zhang, T.; Zhang, H.; Zhang, X.; Deng, Z.; ZhouG. *Nucl. Inst. Meth. Phys. Res.* B 2000, 169, 89-93.
[204] Wu, Y.; Zhang, T.; Zhang, Y.; Zhang, H.; Zhang, X.; ZhouG. *Nucl. Inst. Meth. Phys. Res.* B 2001, 173, 292-298.
[205] Wu, Y.; Zhang, T.; Zhang, Y.; Zhang, H.; Zhang, X.; ZhouG. *Science in China*E 2001, 44, 493-498.
[206] Wu, Y.; Zhang, T.; Liu, A.; Zhou, Gu. *Surf. Coat. Technol.* 2002, 157, 262-266.
[207] Kobayashi, T.; Iwata, T.; Doi, Y.; Iwaki, M. *Nucl. Inst. Meth. Phys. Res.* B 2001, 175-177, 548-553.
[208] Stepanov, A. L.; Abdullin, S. N.; Petukhov, V. Yu., Osin, Yu. N.; Khaibullin, R.I.; Khaibullin, I. B. *Philos. Mag.* B 2000, 80, 23-28.
[209] Stepanov, A. L.; Popok, V. N.; Khaibullin, I. B:; Kreibig, U. *Nucl. Inst. Meth. Phys. Res.* B 2002, 191, 473-477.
[210] Stepanov, A. L. *Tech. Phys.* 2004, 49, 143-153.
[211] Bazarov, V. V., Petukhov, V. Yu., Zhikharev, V. A.; Khaibullin, I. B. *Mater. Res. Soc. Proc.* 1995, 388, 417-422.
[212] Boldyryeva, H.; Umeda, N.; Plaksin, O. A.; Takeda, Y.; KishimotoN. *Surf. Coat. Thechnol.* 2005, 196, 373-377.
[213] Khaibullin, R. I.; Osin, Y. N.; Stepanov, A. L.; Khaibullin, I. B. *Vacuum* 1998, 51, 289-294.
[214] Khaibullin, R. I.; Osin, Y. N.; Stepanov, A. L.; Khaibullin, I. B.*Nucl. Inst. Meth. Phys. Res.* B 1999, 148, 1023-1028.
[215] Stepanov, A. L.; Khaibullin, R. I.; Khaibullin, I. B. Philos. *Mag. Lett.* 1998, 77, 261-266.
[216] Scheunemann, W.; Jäger, H. Z. *Phys.* 1973, 265, 441-454.
[217] Quinten, M. Z. *Phys. B* 196, 101, 211-217.
[218] Khashan, M. A.; Nassif, A. Y. *Opt. Comm.* 2001, 188, 129-139.
[219] Sviridov, D. V. *Russian Chem. Rev.* 2002, 71, 315-327.
[220] Pignataro, B.;Fragala, M. E.; Puglisi, O. *Nucl. Instr. Meth. Phys. Res.* B 1997, 131, 141-148.
[221] Palik, E. D. *Handbook of optical constants of solids*, Academic: London, 1997.
[222] Sinzig, J.; Quinten, M. *Appl. Phys. A* 1994, 58, 157-162.
[223] Wang, H.; Wong, S. P.; Cheung, W. Y.; Ke, N.; Chiah, M. F.; Liu, H.; Zhang, X. X. *J. Appl. Phys.* 2000, 88, 4919-4921.

[224] Ivanov-Omslii, V. I.; Tolmatchev, A. V.; Yastrebov, S. G. *Philos. Mag. B* 1996, 73, 715-722.

[225] Biederman, H.; Chmel, Z.; Fejfar, A.; Misina, M.; Pesicka, J. *Vacuum* 1990, 40, 377-381.

[226] Stenzel, O.; Kupfer, H.; Pfeifer, T.; Lebedev, A.; Schulze, S. *Opt. Mater.* 2000, 15, 159-165.

[227] Kreibig, U.; Gartz, M.; Hilger, A. Ber. Bunssenges. *Phys. Chem.* 1997, 101, 1593-1601.

[228] Hölzl, J.; Schulte, F.; Wagner, H. *Solid surface physics*, Springer: Berlin, 1979.

[229] Pinchuk, A.; Kreibug, U. *New J. Phys.* 2003, 5, 151.1-151.15.

[230] Gartz, M.; Keutgen, C.; Kuenneke, S.; Kreibig, U. *Eur. Phys. J. D* 1999, 9, 127-133.

INDEX

A

B

C

D

E

F

G

H

I

L

M

N

O

P

R

S

T

U

V

W

X

Z